PRODUCTION AND USE OF INDUSTRIAL ROBOTS

UNITED NATIONS
New York, 1985

ECE/ENG.AUT/15

UNITED NATIONS PUBLICATION

Sales No. E.84.II.E.33

02500P

ISBN 92-1-116316-1

CONTENTS

LIST OF TABLES

Page

Page

LIST OF FIGURES

I. INTRODUCTION

The second session of the Economic Commission for Europe (EÇE) Working Party on Engineering Industries and Automation, in February 1982, reviewed past activities and the proposed programme of work in the field of automation. It was suggested that "further assessment of automation trends should be undertaken using selected model topics. In that connection, it was proposed that efforts should be concentrated on manufacturing, including assembly and handling processes in engineering industries, and preferably on flexible automation equipment, such as industrial robots. The importance of precise definition and classification of robots was stressed, namely the necessity to differentiate between manipulators and programmable computer-assisted robots. During the discussion, the Working Party suggested that a study should be undertaken on the subject" (ECE/ENG.AUT/6, para. 14 and annex).

The draft of part one of the study on the production and use of industrial robots, concerning the trends in manufacture and use of industrial robots, was discussed at the third session of the Working Party, in February 1983 (ECE/ENG.AUT/9, paras. 17-20). It was decided to finalize part one of the study and issue it separately, before the information contained therein could become obsolete. Part one was issued as document ECE/ENG.AUT/12 in August 1983.

Concerning part two of the study, on the diffusion of robots in the ECE region, the Working Party, at its third session, decided that it should be based on national contributions, received in response to two questionnaires issued by the secretariat (see annexes I and II to the present study). The 2nd ad hoc meeting for the study, held in November 1983, was informed that 12 ECE member countries had provided data on the 1982 distribution of robots in their countries (ENG.AUT/AC.5/4). Additional verbal information had been provided by the authorities of Sweden and the Union of Soviet Socialist Republics, as well as by the Council of Mutual Economic Assistance (CMEA).

In view of the interest demonstrated in part one of the study, the fourth session of the Working Party (ECE/ENG.AUT/14, paras. 20, 21), held in March 1984, decided to:

(a) Update, where necessary, the text of part one of the study and to issue it, together with part two, in the form of a sales publication. In this respect, the secretariat was asked to contact member countries and international organizations active in the field, requesting them to provide any data for the year 1982 which had so far not been received, as well as any (relevant) new information;

(b) Continue the regular collection of statistical data in simplified form on the diffusion of robots. In this connection the secretariat was asked to investigate the possibility of using the regular questionnaire for the Annual Review of Engineering Industries and Automation as a basis for the collection of the required data until such time as the relevant items on industrial robots had been included in regular statistical nomenclatures such as the International Standard Industrial Classification of All Economic Activities (ISIC) and the Standard International Trade Classification (SITC), etc.; and

(c) Study the subject of trends in the production and use of industrial robots on a regular basis - for example, every two or three years - with a view to undertaking a thorough evaluation of the techno-economic and social impact of robotization on national economies.

The 6th <u>ad hoc</u> Meeting on Questions of Statistics concerning Engineering Industries and Automation, held in July 1984 (ENG.AUT/AC.1/8), recommended that:

- As a first trial, the next questionnaire for the Annual Review should include a request for data on production, imports and exports of industrial robots for 1983 and 1984 (see annex III);

and

- Countries should be encouraged to collect these special statistics from all available national information sources. It was agreed that, when publishing unofficial data transmitted by countries, the secretariat would include a clear indication of their source (e.g. national robot association, scientific and technical society dealing with robotics, etc.).

As a follow-up to the present study, an ECE Seminar "Industrial Robotics '86 - International Experience, Developments and Applications" will be held at Brno (Czechoslovakia) from 24-28 February 1986. Participants will also have an opportunity to visit the fifth International Exhibition of Industrial Robots "ROBOT '86" organized at the Brno Fairgrounds. The provisional programme for the Seminar includes consideration of:

- Recent developments in robots and their components;

- The economic and social impact of the use of robots;

- Governmental and other support programmes and international co-operation;

- Implementation experience in various application areas.

II. SCOPE OF THE STUDY

The <u>ad hoc</u> Meeting on Methodology for Assessing the Techno-Economic Trends of Automation held in November 1981 emphasized the role of robots and the impact of robotization on national economies (ENG.AUT/AC.3/2). Innovation of automation gives rise to constant changes in the structure of manufacturing technologies: the introduction of microelectronics is influencing the concepts of control systems which create the logical and intelligence of automated equipment. In view of this, the appropriate definition and classification of robots is considered a priority in the present study. A relevant and simple classification system also appears to be a prerequisite for the introduction of robots into statistical nomenclatures for a regular collection of data by ECE.

The present impact of computerization and microelectronics was stressed in a note by the secretariat 1/ put before the above-mentioned meeting. It was agreed that robotics seemed to be the most rapidly developing area in the automation of manufacturing processes characterized by a wide range of fields of application. Stress was also placed on the man-machine interfaces in this field and on the fact that the implementation and maintenance of robots required a considerable amount of "teach-in" procedures and sophisticated software, including re-programming procedures.

The rapid advances in mocroelectronics between 1968 and 1988 are indicated in table 1 below.

<u>Table 1.</u> <u>Growth of world electronics and semiconductor markets (1968-1988)</u>

Year	World Electronic market (billions of US dollars)	World semiconductor consumption (billions of US dollars)	Semiconductor share of electronic equipment value (percentage)
1968	35	2	5.5
1978	145	10	7.1
1988 a/	500	50	10.0

<u>Source</u>: <u>The International Microelectronic Challenge</u> (The American Response by the Industry, the Universities and the Government), The Semiconductor Industry Association, Cupertino, California, May 1981.

a/ Rough estimate.

____

1/ Draft proposal for improving the methodology needed to assess the techno-economic trends of automation (ENG.AUT/AC.3/R.1).

Table 2 below illustrates the distribution of electronics industries by main geographical areas.

Table 2. Distribution of world electronics production by geographical areas and different industries (1980)

(Percentage)

Geographical area	Total production of electronics industries in 1980 (percentage)	Share of different industries (Percentage)				
		Computing and office equipment	Communications equipment	Industrial automation and medical electronics	Electronic components and parts	Consumer electronics
United States	38.2	9.6	12.3	4.5	10.0	1.8
Japan	14.5	2.2	1.4	1.4	4.5	5.0
Western Europe	32.8	7.3	10.5	5.9	5.9	3.2
Rest of the world	14.5	0.9	4.5	0.9	2.7	5.5
Total	100.0	20.0	28.7	12.7	23.1	15.5

Source: ECE secretariat estimate based on data provided by the Central Statistical Office of Sweden (SCB).

Similar information for selected European countries is given in table 3 below.

In 1979 the European Community (EEC) began undertaking special research in the field of information technology. 2/ This activity was followed in 1982 by the European strategic programme for research and development in information technology (ESPRIT). A larger scale ESPRIT – a 10-year programme – was launched in 1984 with a budget of approximately 1,300 million United States dollars for the first five years. The following are the five priority areas of ESPRIT activities:

- Advanced microelectronics components and parts;

- Advanced information processing, including man-machine communication;

- New software technology;

- Office automation, including word and image processing and machine translation; and

- Factory automation, including robotics.

2/ The European Community and new technologies – from ESPRIT to the "Biosociety". European File No. 8, Brussels, April 1984.

Table 3. Distribution of electronics production in selected countries (1982)

Country	Approximate value of micro-electronics production 1982 (thousands of millions of 1982 US dollars)	Micro-electronics production as share of GNP 1982 (percentage)	Average annual growth 1978–1982 (percentage)	Share of different industries (percentage)				
				Computing and office equipment	Communications equipment	Industrial automation and medical electronics	Electronic components and parts	Consumer electronics
Germany, Federal Republic of	19.8	3.0	−1	22	21	22	19	15
United Kingdom	14.8	3.1	−9	22	37	19	16	6
France	13.5	2.5	−4	23	42	12	16	6
Italy	6.3	1.8	−17	38	28	12	14	9
Netherlands	4.4	3.2	+19	27	26	18	24	6
Switzerland	3.0	3.1	+33	9	19	32	13	27
Sweden	2.9	2.9	+19	15	46	24	11	4
Belgium	2.1	2.5	−5	21	33	12	16	18
Spain	1.7	0.9	−16	13	34	6	13	36
Austria	1.7	2.5	+29	7	12	10	24	47
Denmark	1.0	1.7	+7	11	24	42	15	9
Finland	0.6	1.3	−3	13	26	22	18	20
Norway	0.6	1.1	−13	19	43	25	10	3

Source: ECE secretariat estimate based on data provided by the Central Statistical Office of Sweden (SCB).

The introduction of microelectronics into robot control has significantly improved the flexibility and reliability of the robot operation itself and also the synchronization and co-ordination of robots with other types of automation, such as numerically controlled (NC) machine tools, automated handling equipment and computer-aided design (CAD). Applications for semiconductor technology have greatly expanded, with annual market-growth rates averaging 25 per cent. As an example of advancing technological sophistication, one can compare the five or so logic circuits placed on a semiconductor device in the late 1960s with the 100 circuits per device in the mid-1970s and the 1,500 circuits per device in 1983. At the same time, the number of bits per semiconductor chip has increased from 1,000 in 1970 to 256,000 in 1980 (the forecast for 1985 is 1 million bits) and the price per bit in United States cents has decreased from 0.1 in 1970 to 0.005 in 1980 (the forecast for 1985 is 0.002 United States cents per bit). 3/

At the level of computer control of engineering processes, the complexity and integrity of particular elements creating the whole automated manufacturing system should be taken into account. An illustrative description of such interactions can be seen in figure 1.

It is obvious that full use of the advantages of robots can be obtained through integration of their operation with other computer-aided manufacturing (CAM) and CAD functions. However, today the majority of robots are implemented at separate work-places or work-stations; for instance there are already a significant number of robots co-ordinated with NC-machine tools (e.g. selection of tools) and automated handling systems. This kind of co-ordination is sometimes characterized as a flexible manufacturing system (FMS) or integrated manufacturing system (IMS). FMS is commonly understood to consist of several machine tools, other NC-appliances and operation modules linked together by a handling system and controlled by a computer-based system. The system is flexible in that it can process different parts at different volumes. Some examples of FMS already implemented were presented at the Conference on Flexible Manufacturing Systems, held at Milan in March 1982. 4/ Programmable robots seem to be an optimum solution for FMS; however with more flexible equipment, production costs increase. The functional relation between flexibility and productivity is indicated in figure 2. 5/ During the last 2 to 3 years the number of events concerning FMS has grown significantly. 6/

3/ The National Swedish Industrial Board.

4/ Fifth NC Industrial Automation and Robot Exhibition, Milan, 1-5 March 1982 (Proceedings of the Conference).

5/ T. Palazzo, "Maximizing Production – The Answer: FMS". Paper presented at the above-mentioned conference.

6/ As from 1982, the IFS (Conferences)Ltd., Bedford (United Kingdom) has been organizing regular international conferences on FMS, the third and most recent being held at Boeblingen (Federal Republic of Germany) in September 1984 (jointly sponsored by the Fraunhofer Institute für Produktionstechnik Automatisierung – IPA Stuttgart.

An interesting forecast was published recently, 7/ stating that by 1990 some 40 to 50 per cent of engineering manufacture in leading industrialized countries would be done on flexible manufacturing systems. A study on the current impact of microelectronics on productivity and the structure of jobs 8/ gave fundamental importance to flexibility:

"Until now, automation has been largely restricted to factories that turn out thousands of identical products, because it has been too costly to retool machines at frequent intervals to perform new tasks. But the development of reprogrammable machinery makes it economically feasible to automate production processes that involve short production runs and frequent changes in machine settings. The majority of manufacturing processes fall into this category."

The 1982 estimates of Creative Strategies International, shown in table 4, may be used to illustrate the relatively faster growth of markets for robots, in comparison with other components of factory automation, such as CAD, CAM and FMS.

Table 4. Industrial automation – estimated value of shipments (1990)

(Millions of US dollars)

Shipments	Year		Average annual growth rate 1981–1990 (percentage)
	1981	1990 a/	
Robotics	12 839	67 150.7	47
CAD	3 749.5	41 745.5	29.2
CAM	1 885	20 965	14.7
FMS	693.6	7 720.8	5.4
Total	12 839	142 905	30.4

Source: G. Schumann, "The Macro- and Microeconomic Social Impact of Advanced Computer Technology". Futures, June 1984, (based on the data from Creative Strategies International).

a/ Estimated figures.

As the present study concerns mainly the problem of robot classification, which is to some extent independent of the nature of surrounding processes, the environment and so on, associated problems such as the solution of IMS, FMS, NC and CNC-machine tools, CAD/CAM, and automated handling are not described. Many of these topics have already been investigated at ECE Seminars, e.g.:

7/ FMS Update, vol. 1, No. 6. IFS Bedford, June 1983.

8/ C. Norman, "Micro-electronics at Work: Productivity and Jobs in the World Economy". Worldwatch paper No. 39, 1980.

(a) Automated Integrated Production Systems in Mechanical Engineering, Prague (Czechoslovakia), November 1976 (see AUTOMAT/SEM.4/2);

(b) Industrial Robots and Programmable Logical Controllers, Copenhagen (Denmark), September 1977 (see AUTOMAT/SEM.5/3);

(c) Computer-Aided Design as an Integrated Part of Industrial Production, Geneva (Switzerland), May 1979 (see AUTOMAT/SEM.6/2);

(d) Innovation in Engineering Industries: Techno-economic Aspects of Fabrication Processes and Quality Control, Turin (Italy), June 1980 (see ENGIN/SEM.6/3);

(e) Automation of Welding, Kiev (Ukrainian SSR), October 1980 (see AUTOMAT/SEM.7/3);

(f) Automation of Assembly in Engineering Industries, Geneva (Switzerland), September 1981 (see AUTOMAT/SEM.8/3); and

(g) Flexible Manufacturing systems: Design and Applications, Sofia (Bulgaria), September 1984 (see ENG.AUT/SEM.3/4).

Some conclusions reached at Seminars referred to in (b), (f) and (g) above and mentioned below are particularly relevant to the present study.

Concerning the Seminar on industrial robots and programmable logical controllers, some inferences, justifying the need for increased robotization, have subsequently been fully proved in practice, e.g.

"- Flexible automation tools and namely robots fill an important gap gap between low cost automation on the one hand and integrated automation systems on the other. They permit the more rational use of both existing equipment and available manpower skills in an evolutionary way, permitting gradual adjustments to the structural changes of the production process and the product pattern, as these evolve in a changing economic situation;

- The important economic and social questions of how far and how fast this technical development is proceeding, have to be resolved so that a balanced economic growth can be ensured. To this effect, the innovation process of flexible automation should proceed as a system approach with producers, users and operators all contributing in a way conducive to their mutual benefit; and

- Special emphasis should be given to defining international standards for industrial robots and programmable logical controllers as regards their operation, and technical parameters and guidelines for their introduction into the production process should be properly formulated (terminology definitions)."

At the Seminar on Automation of Assembly in Engineering Industries, a special section was devoted to the use of robots in automated assembly.

"During the [seminar] discussions, special importance was given to various aspects of the economic and technical justification of robots. The use of industrial robots seemed to be a step in the right direction in solving the problem of automated assembly of engineering products with continuously increasing sophistication, manufactured in larger series and at high manufacturing rates. Although the current pilot installations were still very costly, it was believed that, with the introduction of microelectronics for control functions of both universal and special-purpose robots, their price would decrease.

Another recurring theme was the contradiction arising from the use of flexible robots with rigid assembly equipment, including various peripheral devices, transport equipment and tooling. In this connection, problems such as minimizing the time needed for the resetting of assembly lines and work-stations, reducing the space required for the fulfilment of different auxiliary operations, etc., were also discussed. The exchange of information on solutions in use in ECE member countries in this field, including the standardization of specific units and modules, would be extremely useful.

Participants stressed that, with the introduction of new hardware, the problem arose of adequate development of the software. Currently, manufacturers and research institutes were developing their own specialized programming languages, leaving aside such important features as compatibility and portability. These features were required not only for implementation of assembly systems composed from units supplied by various manufacturers, but also for their integration with the automation equipment used for production control and data processing. Namely, the higher-level universal programming languages would certainly be used in most applications in the near future."

The Seminar on Flexible Manufacturing Systems: Design and Applications drew attention to the extensive progress made during the last 2 to 3 years in the field of manufacturing automation in most ECE member countries. Industrial robots together with NC-machine tools and automated handling equipment (including automated guided vehicles) create the technical basis for the introduction of FMS, and many conclusions and recommendations discussed during the Seminar are valid for all robot-based manufacturing systems, e.g.

"- During the discussions, the importance of system approach towards the implementation of FMS was underlined, taking into account not only the hardware solutions, but also the related organizational, economic and human aspects. Special attention should be given to adequate verification and testing of the algorithms and computer programs used. It was pointed out that the proper functioning of the FMS was very much correlated to the material requirement planning and other related production control and scheduling operations;

"- For reliable functioning of FMS it was also important to design an optimum structure for the computer control system, including the solution of all necessary interfaces (man, machine, computer), Sophistication of the control system used certainly depended on the availability of resources (both for hardware and software). The control of larger FMS installations should, wherever possible, be based on a hierarchical decentralized computer system;

"- Special attention was given to the use of modular robotic systems and
    a brief overview of major trends in the development of kinematics of
    advanced industrial robots and manipulators from the point of view of
    their effective use in FMS was presented. Industrial robots were
    analysed with particular focus on robots with adaptive capabilities
    as an important prerequisite to the development of advanced FMS
    structures. Special consideration was given to the modular design
    and construction of industrial robots and multi-machine systems;

"- It was pointed out that systems currently used for assembly were
    designed primarily for large-batch production in the electrotechnical,
    automotive and similar industries. When dealing with the assembly of
    small batch sizes, the problem arose of orientation, position and
    recognition of parts and components to be assembled. Sensor systems
    provided a solution to those requirements. Such systems should be
    equipped with multi-sensor capabilities. A further problem
    consisted of the design of flexible grippers permitting the different
    components to be handled effectively. It was emphasized that, in
    order to achieve efficient flexible assembly, it was necessary to
    design the product in such a way as to facilitate its automated
    flexible assembly;

"- During the discussion concerning the standardization of FMS and its
    components (particularly in the field of manipulation equipment and
    robotics) the need for improved international co-operation was
    emphasized. Some examples of approaches used in Japan were provided
    (standardization of pallets). In this connection, participants were
    informed about the creation of a new International Organization for
    Standardization (ISO) Technical Committee - TC 184 - dealing with
    industrial automation. It was explained that that committee had
    five working groups, including a working group on industrial robots,
    which had already started their activities; and

"- Information was provided on standards of the International
    Electrotechnical Commission (IEC) related to the manufacture and
    testing of robots and their use. A retrospective summary of the
    work of IEC and a description of its methods of work was presented,
    emphasizing that in future national and regional standards would be
    replaced to a greater extent than at present by international
    standardization."

The ECE Working Party on Engineering Industries and Automation, has
included in its current programme of work, in addition to the regular study of
the production and use of industrial robots, the study of other topics closely
related to automation (see ECE/ENG.AUT/14, annex I):

- The prospective medium- and long-term impact of automation on the economy
   as a whole and on individual branches in particular;

- Study on the methodology, needed, to assess, the techno-economic trends
   and the level of automation;

- Recent developments in telematics equipment and the application of
   information technology;

   Review of recent technological trends in electrical and electronic
   engineering;

- Recent developments in software means for industrial automation; and

- Recent trends in flexible manufacturing.

The number of international events planned on automation-oriented topics is growing significantly. These are being organized either by international non-governmental organizations active in the area of automation (e.g. IFAC, IFIP, IFORS, IMEKO, etc.) or by specialized national societies, institutes, agencies, etc. For example, the Computer and Automated Systems Association (CASA) of the Society of Manufacturing Engineers (SME) has established a tradition of so-called AUTOFACT (Automated Factory) Conferences and Exhibitions, which are organized in United States, Europe and Japan, in order to:

- Provide professionals with updated information on many important aspects of computer-based automation of manufacturing;

- Provide liaison among industry, government and education in identifying areas for further development and introduction of new manufacturing technologies; and

- Encourage the design and application of fully integrated manufacturing facilities.

The first AUTOFACT EUROPE Conference and Exhibition was held at Geneva in September 1983 and the second at Basel in September 1984. The development in robotics remain one of the main AUTOFACT themes.

The introduction of robots into manufacturing processes seems to be a promising solution to current problems such as:

- Technological innovation allowing higher production output, including high-speed and continuous manufacturing (two or three shifts), improvement of quality and reliability of manufacturing process;

- Increase of the economic efficiency of production, including higher capital productivity, material savings, lower labour costs etc. It is expected that an average programmable robot may save between two and three skilled workers from having to carry out simple repetitive operations; and

- Various social problems such as the replacement of workers and operators in hazardous and physically and mentally demanding and tiring jobs, allowing their transfer to more convenient and skilled jobs (in Japan, about 1 million workers and operators annually suffer from work-related injuries and illnesses). 9/

Any further analysis of the technical, economic and social impact of robotization requires the collection of relevant data concerning production and international trade in robots and their distribution by various types, operational modes and application fields (see chapters III and IV). The present study therefore reviews general trends in the production and use of industrial robots and proposes a possible approach towards future work in this field within the Economic Commission for Europe.

9/ "Robots in the Japanese Economy", ed. K. Sadamoto. Survey Japan, Tokyo 1981.

III. ROBOT DEFINITION AND CLASSIFICATION

The first attempts to build an automat were made in ancient Greece 2,300 years ago by Archytas of Tarent and, later, by Heron of Alexandria. The first androids, i.e. automats with human looks, were built during the Renaissance by Leonardo da Vinci; later, in the eighteenth century, such machines were constructed by, among others, J. de Vaucanson in France and P.J. Droz and Lisson in Switzerland. Although the first artificial human beings had been described earlier (e.g. Goethe: Faust), the term "robot" was introduced only in 1920. 1/ The first robot-like equipment, Televox, was built in 1927 by R.J. Wensley (Westinghouse, Pittsburg). The world's first industrial robot went into service in 1961 (General Motors, Trenton). 2/

In the 1960s, robots were introduced for simple manipulations, in the 1970s, with improvements in computer technology, programmable robots were developed, and in the present decade, robots with sensory perception (third-generation intelligent robots) are coming into use.

The following definition of an industrial robot was proposed by the International Organization for Standardization (ISO): 3/

"The industrial robot is an automatic position-controlled reprogrammable, 4/ multi-functional manipulator having several degrees of freedom capable of handling materials, parts, tools, or specialized devices through variable programmed motions for the performance of a variety of tasks. 5/ ... It often has the appearance of one or several arms ending in a wrist. Its control unit uses a memorizing device and sometimes it can use sensing and adaptation appliances that take account of environment and circumstances. These multi-purpose machines are generally designed to carry out repetitive functions and can be adapted to other functions without permanent alteration of the equipment."

Very close to the ISO definition is that used by the British Robot Association 6/ which, in simplified form, is as follows:

1/ K. Capek, _Rossum's Universal Robots_. Prague 1920 (robota = hard work, compulsory service, etc.).

2/ P. Marsh, "America's factories race to automation". _New Scientist_, 25 June 1981.

3/ "Work plan: Industrial Robots"; ISO/TC 97/SC 8 N, November 1980 (prepared by AFNOR — Association française de normalisation).

4/ The term "reprogrammable" means that the mechanical arm is computer-controlled to qualify as a robot.

5/ This definition is nearly identical to that of The Robot Institute of America (RIA)

6/ _Robotfacts_. British Robot Association, December 1982.

"An industrial robot is a reprogrammable device designed to both manipulate and transport parts, tools or specialized manufacturing implements through variable programmed motions for the performance or specific manufacturing tasks."

A more general definition (covering also "non-industrial" robots) underlining software control is used in the Penguin dictionary: 7/

"Any device equipped with sensors capable of detecting input signals and environmental conditions and with reacting and guidance mechanisms; capable of performing calculations on the input data in accordance with a stored programme and consequently able to run itself".

Webster's dictionary 8/ introduces the human element into the definition:

"An automatic apparatus or device that performs functions ordinarily ascribed to human beings or operates with what appears to be almost human intelligence."

According to Japanese sources 9/ a robot is defined as:

"A mechanical system which has flexible motion functions analogous to the motion functions of living organisms or combines such motion functions with intelligent functions, and which acts in response to the human will. In this context, intelligent functions mean the ability to perform at least one of the following: judgement, recognition, adaptation or learning."

The first ad hoc Meeting for the study on Production and Use of Industrial Robots held in November 1982 10/ accepted the robot definition as proposed by ISO.

Some authors, among them I.M. Havel, 11/ underline the importance of such aspects as the interface with information-processing, decision-making and feedback functions. From this viewpoint, the robot system consists of three sub-systems (fig. 3):

- Cognitive sub-system or control system;
- Motoric sub-system or system of moving parts; and
- Sensoric sub-system. 12/

The system of moving parts (mechanical structure) includes the arm (articulated part of the robot), the wrist (end part of the arm circulating the tool), the gripper (end effector for grasping and holding) and the hand (part of the gripper

---

7/ The Penguin Dictionary of Microprocessors. Penguin Books Ltd., 1981.

8/ As cited in J.F. Engelberger, Robotics in Practice. Kogan Page, London, 1980.

9/ Japanese Industrial Standards, JIS B 0134 (1979).

10/ Report of the ad hoc Meeting (ENG.AUT/AC.5/2).

11/ I.M. Havel, Robotika – Introduction to the Theory of Cognitive Robots. SNTL Publishing House, Prague 1980 (in Czech).

12/ See also J.F. Engelberger, Robotics in Practice. Kogan Page, London,1980.

consisting of jaws capable of grasping). The control system varies from a simple electro-mechanical system based on the use of fixed stops and limit switches to a sophisticated computer control with on-line reprogrammable memories. According to the sophistication of the control system, robots can be divided into the following groups:

- Limited sequence robots;
- Robots with point-to-point control;
- Robots with continuous path control;
- Intelligent robots with sensory perception.

Leaving aside manually-operated robots-manipulators (pick and place), the various groups or robots can be broken down as follows: 13/

- A <u>fixed-sequence robot</u> which performs successive operations in a pre-determined unchangeable sequence and a <u>variable-sequence robot</u> which will accept a change in the order of the operations;

- A <u>point-to-point</u> playback robot which uses the memory device to execute a set of operations that were originally performed under manual control;

- A robot with <u>continuous path control</u> which is already programmed off-line or on-line and which may be taught in real time (the operator teaches the robot by leading the arm following the required path, speed etc.); and

- An <u>intelligent robot</u> which uses sensory perception (video-sensors, tactile-sensors etc.) to detect changes in working conditions and reacts accordingly, using its programmed decision-making capabilities.

The following comparison from the programming viewpoint shows the considerable advantages of continuous path control. 14/

<u>"Point-to-point control</u>	<u>Continuous path control</u>
It is not possible to use a Cartesian movement procedure with constant retention of tool orientation when setting up.	It is possible to use a Cartesian movement procedure with constant retention of tool orientation when setting up.
Each individual point has to be adjusted by approaching and storing; the point spacing must, for example, be less than approximately 15 mm in order to avoid inaccuracies.	In the case of straight lines, only the starting point and the end point have to be approached and stored, regardless of the length.
The tool orientation has to be adjusted at each point.	The tool adjustment has to be set only once. It is automatically kept constant or linear interpolated by the control.

13/ "A report on robotics in Japan", <u>Robotics Today</u>, No. 3 (1981).

14/ H. Wörn "A new Generation of Robot Controls". Conference on Robots in the Automotive Industry, Birmingham, United Kingdom, April 1982 (see also Annex V).

Adjustment of the tool orientation is not independent of the tool centre point position.

Adjustment of the tool orientation is independent of the tool centre point position."

According to a French source 15/ robots can be divided into three categories:

- Pick and place robots (small manipulators) for handling parts – approximate price between $US 1,000 and 3,000 for light loads and from $3,000 to 10,000 for heavier loads;

- Middle range robots (large manipulators), with programmable control, also used mainly for handling operations – approximate price between $30,000 and 40,000; and

- Top range robots – priced at around $100,000 – capable of different handling and machining operations, various assembly tasks – completely reprogrammable and linkable with TV recognition equipment.

A slightly different classification of robots 16/ includes intelligent robots (third-generation robots):

- <u>Robots for simple manipulations</u> (without servocontrol) which are used as a linkage between specialized manufacturing and handling machines. These work at a fixed rhythm, with point-to-point control and are usually fitted with a hydraulic or pneumatic power unit. They are less flexible but very reliable;

- <u>Programmable robots</u> (with servocontrol) which are already fitted with sensors for three-dimensional control and error-correction. They are based on continuous path control and are fitted with memory devices (several programmes), which make them very flexible in application; and

- <u>Intelligent robots</u> which are programmable robots, able to modify their own functioning. They are fitted with sophisticated sensor-systems (optical, tactile, voice). They are still being tested in laboratories and are relatively expensive (e.g. special software is needed to change sensor signals into precise mechanical movements).

However important the nature of the control system, the physical capability of the robot is still of primary importance; it includes:

- Arm geometry;
- Drive system;
- Dynamic performance and accuracy;
- Reliability and safety.

15/ <u>Robot News International</u>, vol. 2, No. 6, March 1982.

16/ Mackintosh Consultants Co. Ltd., Luton 1979.

Concerning arm geometry, there are a variety of geometric configurations (co-ordinates) defining the movement of the robot arm, wrist and end effector. 17/ These include:

- In the <u>Cartesian system</u> (three axes) the gripper moves in three different directions controlling the height, the width and the depth of the operation. It gives the robot's arm great accuracy but makes it slow;

- The <u>cylindrical system</u> (two axes and one angle) consists of an extendable arm and a central pole on which it is mounted. The arm can move horizontally out from the pole, swivel round it and move vertically up and down along it;

- The <u>polar system</u> (one axis and two angles) consists of an extendable arm mounted on a central pivot. Like a cylindrical arm, a polar arm can swivel round its mounting. Instead of moving up and down it, however, the arm tilts to reach out above or below the level at which it is mounted; and

- The <u>revolute (anthropomorphic) system</u> (three angles), in which human-like arms can bend and swivel at the "shoulder" and bend at the "elbow". These are the most flexible of robotic arms, capable of reaching nooks and crannies that others cannot.

In addition, any of these arm co-ordinate systems requires three articulations to deliver the wrist anywhere in the sphere of operation and three more articulations for the orientation of the end effector. Usually, depending on the type of work, the geometry of the workpiece etc., the robot is able to work with less than six articulations. The total number of axes is sometimes identified as the number of degrees of freedom which is considered an indication of the degree of versatility. Another indicator of the degree of versatility is the volume of the operation space (robot's working space).

The movable parts of the robot are driven by the following kinds of devices: 18/

- <u>Pneumatic drives</u> which use compressed air to move the robot's mechanical arm. These are lightweight, fast and relatively inexpensive. Pneumatically driven robots typically sell for $8,000-20,000 apiece. But they cannot provide much strength;

- <u>Hydraulic drives</u> which use compressed fluids to drive the arm. Although hydraulically driven robots are more expensive than pneumatically driven ones, they are much stronger. One problem, some robot-makers say, is that they are prone to messy leaking; and

---

17/ "Robots are coming to industry's service", <u>The Economist</u>, 29 August 1981.

18/ <u>Ibid</u>.

- <u>Electric motors</u> which are the strongest and least energy consuming of
  robot drives; they are also the most expensive, both to buy and to
  maintain. Electrically driven robots cost at least $80,000.

While the drive will make the robot's arm move, there must also be a way to make
it stop. This can be done in one of two ways: by simply cutting off the drive -
and slamming on "brakes" - once the robot's arm has reached a desired position;
or, alternatively, by using a more expensive servocontrolled system which can
begin to slow the arm down as it nears the desired stopping point and then using
the drive's own power to bring it to rest. Servocontrolled systems can even
reverse the drive to the arm if it overshoots its mark. According to
J.F. Engelberger, 19/ the most commonly used drive form is the hydraulic
(around 50 per cent) because of its power, accuracy and compactness. About
30 per cent of today's robots use pneumatic systems (low operating and maintenance
costs, availability of compressed-air lines) and the remaining 20 per cent use
electro-mechanical drives (servomotors, stepping motors, etc.).

Experience has shown that robots with a more dynamic performance are less
accurate. It is quite logical that, with a heavier load (moment of inertia),
a higher operational speed or a larger operating space, it is more complicated
and expensive to attain the required accuracy. Reliability is defined as the
capability of an industrial robot to perform the given task under specified
operating conditions. The main safety requirement is the guarding and protection
of the whole working area and the prevention of unauthorized access. In this
connection, appropriate training of personnel entering the guarded area to perform
their functions (operating, controlling, maintenance etc.) is also required.

A survey of 18 frequently used types of industrial robots was published in
February 1982. 20/ It gives some idea of the capability of today's robots:

- Concerning the load capacity, robots can be roughly divided into three
  groups - those with a capacity of from 5 to 20 kg, those with a capacity
  of up to 100 kg and those with a maximum capacity of 1,000 kg;

- The horizontal and vertical reach which varies from a few centimetres
  to approximately 3 metres;

- The degrees of freedom (number of articulations) which vary from
  two to six (over 50 per cent of the described robots have six
  articulations);

- The fact that 80 per cent of robots are computer controlled (50 per cent
  by microprocessors), with a memory capacity of up to 4 K bytes and
  teach-mode programming; and

19/ J.F. Engelberger, <u>Robotics in Practice</u>. Kogan Page, London 1980.

20/ M.S. Janjua, "Industrial robots survey" Systems International, PIC,
February 1982.

- That nearly 60 per cent of robots use hydraulic drive and nearly
  30 per cent the DC servodrive (including combinations with
  hydraulic and pneumatic control of the same functions).

An analytical study covering 500 models of robots undertaken in the
Soviet Union 21/ provided the following information:

(a) As regards the arm geometry the majority of today's robots use the
cylindrical system (58.5 per cent), but 20 per cent of robots are still
based on the two- or three-dimensional Cartesian system;

(b) Concerning the number of articulations (degrees of freedom), some
63 per cent of robots have four or five articulations (see figure 4);

(c) As for the drive system, 43.4 per cent of models use hydraulic
drive, 43.3 per cent pneumatic drive and already 13.3 per cent are electrically
driven;

(d) Concerning the charateristics of the control system, figure 5, shows
the distribution of robots by the number of programming instructions (stored
in the memory) and figure 6 shows their distribution by input/output
interface channels;

(e) The distribution of robot models by load capacity is illustrated
in figure 7, which shows that some 65 per cent of models are able to
manipulate loads of between 5 and 40 kg. As regards the working space,
75 per cent of models operate in an area of from 0.1 to 10.0 m^3
(see figure 8); and

(f) The accuracy of performance of industrial robots is shown in
figure 9. More than 60 per cent of models could guarantee a positioning
error of less than 1 mm.

Research and development in the field of industrial robots being
undertaken in the USSR is expected to result in the fulfilment of the technical
requirements shown in table 5 below.

____

21/ Based on a contribution by Y.G. Kozyrev, "Perspectives of development
and use of industrial robots", submitted by the Government of the
Union of Soviet Socialist Republics.

Table 5. Technical level of industrial robots planned for production in the Union of Soviet Socialist Republics (1985-1990)

Indicator	Load capacity in kg				
	- 1	2-10	11-160	160-1 000	1 001-
Linear movement velocity in m/s	-1.5	1.6-2.5	1.6-2.0	0.5-1.0	0.5-1.0
Angular movement velocity in grad/s	360	270	120-270	60-120	- 60
Positioning error in mm	-0.01	-0.1	-0.5	0.5-5.0	- 5.0
Memory capacity of simplified robots in K bytes	less than 4				
Memory capacity of multi-functional robots in K bytes	more than 4				

Source: Y.G. Kozyrev "Perspectives of Development and Use of Industrial Robots". Submitted by the Government of the USSR.

Figure 10 shows the interfaces between the type of robot, type of drive, controlling mode and memory devices.

Another approach to the classification of loading and handling manipulators and robots is presented in table 6.

Table 6. Classification of robot-based equipment by type of control

Equipment type / Type of control	Non-programmable devices		Programmable devices				
	Manipulators, tele-operating devices etc.	Simple loading devices	Handling automats	Industrial robot generation			
				1	2	3	4
Adjustable or programmable path	No	Yes	Yes	Yes	Yes	Yes	Yes
Adjustable or programmable sequence of operations	No	No	Yes	Yes	Yes	Yes	Yes
Path control memory type	No No	Mechanical	Mechanical	Electro-mechanical	Electrical	Electrical	Electrical
Switch control memory type	No	No	Mechanical	Electrical	Electrical	Electrical	Electrical
Type of control	No	(Point-to-point)	(Point-to-point)	(Point-to-point)	Mainly multi-point	Mainly continuous path	Continuous path
Sensors characteristics	No	No	No	No	No	Signals generated by tactile sensors	Signals generated by optical sensors

Source: PROCAM Automatisierungstechnik GmbH, Berlin 1982.

Systematic research work in the field of robotics, including classification problems, is being undertaken by the University of Aston at Birmingham, (United Kingdom).   Their approach towards robot classification is shown in table 7.

Table 7. Classification of robotic and related devices

Class of device	Autonomy	Physical versatility	Control versatility
Servo-assisted hoists; teleoperator-telechiric systems for remote handling	None (require human operation)	Varying	Supplied by human operator
Pick and place devices	Yes	Low	Generally low
Playback robots	Yes	Varying: low-high	Low-medium a/
Computer controlled robots	Yes	Varying: Medium-high	High a/
Modular robots	Yes	Medium-high (can be built to required level)	Varying a/
Mobile robots	Yes	Varying: free to move bodily	Varying a/
FMS and other advanced manufacturing systems (may include robotic elements)	Varying	Varying	Varying a/

Source: J. Fleck, "The Adoption of Robots in Industry". Phys. Technol., Vol. 15. The Institute of Physics (United Kingdom), 1984.

a/ Types of robot complying with the ISO definition.

The first ad hoc Meeting for the present study considered the following two contributions submitted by ISO:

- Glossary of terms for industrial robots (draft); and

- Industrial robots - classification and graphic representation (draft).

As both contributions might be used in future for more detailed studies concerning definitions and classification, the parts relevant to the present study have been extracted and are presented in annexes IV and V.

This chapter gives the general classification of robot types while chapter IV covers the classification of robots by operational modes, i.e. robot applications where either the workpiece or the tool is handled by the robot arm. Chapter IX contains a draft proposal for data collection on the diffusion of robots by:

- Type of robot;

- Operational mode; and

- Application field.

The ad hoc Meeting was of the opinion that the classification of robots should answer two main requirements:

- In the long term it should be based on the classification suggested by ISO, i.e. power source, type of motion control, programming methods used and type of sensory interaction; and

- For the purposes of the present study, robots should be divided into two main groups, i.e. programmable robots and robots with sensory perception, in order to simplify the collection of statistical data and information.

The Meeting agreed that the adoption of the above concept did not imply that the automatic manipulators (reprogrammable only by means of hardware) were less important for industrial development; but as they were difficult to count by means of regular statistics, all available sources should be used, even if only in description form, for possible consideration within ECE.

IV. MAIN AREAS OF ROBOT APPLICATION

While chapter III of the present study dealt with the problem of robot definition, a description of robot anatomy, characteristics of robot control etc., in this chapter, which is mainly based on generally available information sources, a brief review is given of the fields of application. This includes the classification of robot applications by:

- Operational modes, i.e. processes where either the workpiece or the tool is handled by robot; and

- Industrial and non-industrial sectors using robots selected by frequency of applications.

When considering total robot population figures, one must take into account the different approaches towards robot definition adopted in various countries and by relevant organizations. As the majority of sources exclude non-adjustable manipulators with manual control when estimating the total robot population, a similar approach was used in the present study.

As regards the areas of application of industrial robots, they can generally be divided into three groups: 1/

- Applications where the workpiece is gripped by robot;

- Applications where the tool is handled by robot;

- Assembly robots.

Some representative processes where robots are successfully used today are described in the following paragraphs, under each area of application mentioned above.

Typical applications where the <u>workpiece is gripped by robot</u> include:

- Handling or transport of workpieces, such as simple handling operations (from a fixed position to another fixed position), or complex transfer including pelletizing, stacking, packing, sorting etc.;

- Loading and unloading of various pieces of manufacturing equipment, such as casting and pressure die-casting, cold or hot pressing, loading and unloading of furnaces (heat treatment operations), injection moulding, soldering, brazing etc.; and

- Manipulation of material within such technological processes as forging, fitting and investment casting.

Typical applications where the <u>tool is handled by robot</u> include:

- Various types of metal working, such as cutting, drilling, grinding and chipping;

1/ R. Zermeno, R. Mosley and E. Braun, "The Robots are Coming – Slowly", <u>The Microelectronics Revolution</u>, edited by T. Forester. Oxford, Basil Blackwell, 1980.

- Joining of materials, e.g. welding (spot, arc or stud), gluing and wiring;

- Surface treatment, e.g. paint spraying, surface coating, powdering, finishing, cleaning and sealing; and

- Auxiliary operations, such as various types of testing, marking and stamping, gathering of products and packaging.

Concerning assembly robots, their use was considered by the ECE Seminar on Automation of Assembly in Engineering Industries (for conclusions of the Seminar see chapter II). There were three main types of assembly robots in use in 1982 in western Europe (i.e. Olivetti/SIGMA, DEA/PRAGMA and Unimation/PUMA) 2/ and experience has shown the following facilities to be essential for robot-based assembly:

- Tactile sensors;

- Optical sensors;

- Versatile gripping devices;

- Assembly oriented fast computer software;

- Special kinematic joints; and

- Sophisticated adaptable feeding devices.

However expensive it is to include even a few of these in the robot system, it seems that the growth of robot applications in assembly processes (1-2 per cent of total robot population in 1982) is strongly correlated to successful development (research, testing etc.) in this field.

Similar views regarding the importance of assembly robots can be implied from the analysis of application areas undertaken in the Union of Soviet Socialist Republics.3/

In another paper presented at the above-mentioned ECE Seminar, the following figures were presented concerning the current situation in Italy (table 8): 4/

2/ H.J. Warnecke and D. Haaf, "Robot in assembly - current developments", paper presented to the UN/ECE Seminar on Automation of Assembly in Engineering Industries, Geneva, 1981 (AUTOMAT/SEM.8/R.20).

3/ "Professions of robots", Socialisticheskaja Industrija, 5 March 1981 (in Russian).

4/ P.U. Bassignana and A. Romiti, "Perspectives and trends in the production of automated assembly machines", paper presented to the UN/ECE Seminar on Automation of Assembly in Engineering Industries, Geneva 1981 (AUTOMAT/SEM.8/R.13).

Table 8. Ratio of number of workers involved in
assembly and machining in Italy

(Percentage)

Ratio of numbers of workers according to their jobs	Industrial sector			
	Automobile	Agricultural machines	Electro-domestic appliances	Light electromechanics
Assembly Total	32	31	23	43
Machining Total	14	20	8.5	12
Assembly Fabrication (fabrication = machining + assembly)	70	61	73	78

Source: V. Gianesini, "La razionalizzazione delle operazioni di assemblaggio", Rivista di Meccanica, No. 704, December 1979.

An analogous situation, according to the same source, exists in the United States (table 9).

Table 9. Share of assembly operations by selected industrial sector in the United States

(Percentage)

Industrial sector	Assembly as percentage of direct labour hours (for assembly + machining)
Transportation	92
Telecommunications	57
All industry	53

Source: K.R. Treer, Automated Assembly. Dearborn, Society of Manufacturing Engineers, 1979.

It seems logical that robots would preferably be introduced into the most labour-consuming stages of engineering manufacturing processes (machine tools, assembly operations etc.). More details on this subject, i.e. labour intensity and labour-distribution of various engineering processes, are given in chapter VIII. The following tables show the distribution of robots by operational modes and by different industrial sectors.

As regards the structure of robot applications, an analysis of available figures for the United Kingdom, the Federal Republic of Germany and the United States is shown in table 10.

Table 10. Distribution of robots by operational mode in the United Kingdom, the Federal Republic of Germany and the United States

(Percentage)

Type of application	United Kingdom			Federal Republic of Germany		United States
	1981	1982	1983[a]	1981	1982	1982
Workpiece gripped by robot:						
Handling and transport	3.7	4.2	4.3	16.2	14.9	31.2
Loading and unloading of equipment	28.4	30.7	31.1	5.9	10.9	39.5
Process manipulation	1.4	0.4	0.5	1.2	1.5	-
Sub-total	33.5	35.3	35.9	23.3	27.3	70.7
Tool handled by robot:						
Metal working	0.8	1.1	1.3	0.4	0.6	0.5
Joining, including welding	35.7	35.2	35.1	44.0	54.7	23.4
Surface treatment	12.3	10.8	10.1	10.0	11.3	4.6
Auxiliary operations	1.0	1.3	1.4	-	-	-
Sub-total	49.8	48.4	47.9	54.4	66.6	28.5
Assembly robots	2.1	2.8	3.1	4.3	3.5	0.8
Other applications (non-industrial, education, research, etc.)	14.6	13.5	13.1	18.0	2.6	-
Total	100.0	100.0	100.0	100.0	100.0	

Source: Robot News International, vol. 2, No. 8, March 1982 (1981 data) British Robot Association, Robotfacts, December 1982 (1982 and 1983 data).

a/ Estimate.

Table 11 shows the distribution of robots by type of application in 1983 compiled on the basis of data provided by national robot societies in France, Finland, the Federal Republic of Germany, Sweden and the United Kingdom.

Table 11. Estimate of robot distribution by type of application in selected countries (1983)

(Units)

Type of application	Federal Republic of Germany	France a/	Sweden	Italy a/	United Kingdom	Finland
Spot Welding	1 560	1 183	375b/	650	349	–
Arc Welding	856	338	b/	150	234	48
Surface Coating	586	236	310	200	167	18
Grinding/Fettling	22	45	–	–	27	12
Assembly	248	262	20	200	103	–
Casting	132	–	210	–	52	–
Machine Tool	320	1 232c/	740c/	–	165	19
Press Tool	121	c/	c/	–	48	–
Inspection/Test	–	45	–	100	30	–
Other Handling	775	–	160	300	364	–
Not Specified	100	125	35	200	133	23
Education/Research	80	125	–	–	81	–
Total	4 800	3 592d/	1 850	1 800	1 753	120

Source: "Europe overtakes USA", The Industrial Robot, Vol. 11, No. 1, March 1984.

a/ Estimated figures.

b/ Includes arc welding.

c/ Includes all loading and unloading.

d/ Includes 1,422 industrial manipulators.

The trends in the distribution of industrial robots by various areas of application in the Union of Soviet Socialist Republics (1980-1985) are shown in table 12.

Table 12. Distribution of industrial robots by main area of application in the Union of Soviet Socialist Republics (1980-1985)

(Percentage)

Area of application	1980	1985 a/	Change
Handling and transport operations	5	4	− 1
Loading and unloading of machine tools	8	12	+ 4
Foundry operations	15	9	− 6
Forging and pressing	16	14	− 2
Heat treatment (furnaces)	7	8	+ 1
Surface treatment (paint spraying, galvanic coating)	30	20	− 10
Welding	5	13	+ 8
Automatic testing	2	3	+ 1
Assembly operations	7	8	+ 1
Others	5	9	+ 4
Total	100	100	0

Source: As for table 5.

a/ Estimate.

Table 13 presents the main areas of application of industrial robots in the German Democratic Republic.

Table 13. <u>Distribution of robots by operational mode in the German Democratic Republic in 1980</u>

(Percentage)

Type of application	Share
Workpiece gripped by robot:	
Loading and unloading of machine tools	20
Metal forming, including forging, investment casting, etc.	8
Loading and unloading of metal pressing equipment	12
Other forming, including die-casting	10
Plastic injection moulding	15
Heat treatment	5
Handling of workpieces, e.g. palletizing and packaging	8
Tool handled by robot:	
Welding	10
Surface treatment including paint spraying	8
Assembly robots	4
Total	100

<u>Source</u>: J. Volmer <u>et al</u>., <u>Industrieroboter</u>. Berlin, German Democratic Republic, VEB Verlag, Technik, 1981 (in German).

Table 14 gives an indication of the trends in the diffusion of robots by various technological applications in Italy.

<u>Table 14</u>. <u>Estimated trends in robot distribution
by area of application in Italy
(1979-1991)</u>

(Percentage)

Application area	1979	1985[a]	1991[a]
Loading and unloading of machine tools	7	8.5	8.3
Forming - metals	7	8.5	9.2
Forming - plastics	3.5	3.2	3.7
Pressing - metals	7	6.4	6.5
Pressing - plastics	7	6.4	3.7
Furnace loading	3.5	5.3	5.6
Heat treatment	5.2	3.2	3.7
Production of consumer goods	8.5	10.6	6.5
Palletizing	5.2	5.3	5.6
Spot welding	13.7	8.5	8.3
Continuous welding	3.5	4.3	4.6
Assembly robots	10	10.6	13
Surface treatment, including painting	8.5	7.5	7.4
Cabling	1.7	2.1	1.8
Measurement, testing	5.2	5.3	5.6
Others	3.5	4.3	6.5
Total	100.0	100.0	100.0

<u>Source</u>: P. Vicentini. "I robot nell' industria dell 'automobile: una breve panoramica" in <u>Notiziaro Tecnico AMMA</u>, No. 10, Torino 1981 (Speciale robotica).

<u>a</u>/ Estimate.

These figures are evidence of the increasing share of assembly robot applications, while the share of surface treatment (paint spraying) and spot welding, both mainly in the automotive industry, will probably decrease.

According to another source, 5/ typical robot operations can be grouped into four categories:

____

5/ W.B. Heginbotham, "Robots and Automatic Factories", <u>British Business</u>, 24 April 1981.

" - Category **A**: Operations such as paint spraying, shot peening, spot welding, stud welding, flame cutting, applying gasket sealing compound, coating, marking, heat sealing, glass cutting, water jet cleaning, drilling and any routines where <u>no gripping</u> is required. With a simple code identification signal, some robots can store a multiplicity of pre-recorded programmes and so cope with a mixed component requirement.

   - Category B: <u>Low-precision gripping</u> operations where standard gripping surfaces are available, for instance gripping surfaces on the component or standard job holders, unloading die casting machines, unloading injection moulding machines, dipping, quenching, handling, investment castings, unloading presses, unloading furnaces, flat sheet transfer, brick handling and stacking, heat treatment, rough transfer operations, stacking pallets and sacks, loading and unloading machine tools over a range of cylindrical or other regular prismatic workpieces;

   - Category C: Where more precise operations are required necessitating interstage <u>tooling and special gripping</u> for each different operation, transfer between stations for hot and cold forging operations, transfer between metal stamping or drawing operations, loading and unloading machine tools on general work inspection, probing, filament winding, wire wrapping, sprue cutting, sorting and packaging, deburring, drilling, routing and arc welding. The needs of any particular operation will have to be carefully scrutinized in order to select appropriate activities;

   - Category D: <u>Assembly</u> and parts control within a factory where variability levels are high."

An analysis of 600 typical robot applications for present-generation industrial-robot devices in 1980 in the Federal Republic of Germany and 371 installations in 1981 in the United Kingdom shows that the percentage of activities in the above-mentioned categories are as shown in table 15.

<u>Table 15</u>. <u>Distribution of robots by category in the</u>
<u>Federal Republic of Germany (1980) and</u>
<u>the United Kingdom (1981)</u>

(Percentage)

Category	Federal Republic of Germany	United Kingdom
A (No gripping)	45.0	36.0
B (Low-precision gripping)	42.0	48.0
C (Tooling and special gripping)	12.5	15.0
D (Assembly)	0.5	1.0
Total	100.0	100.0

<u>Source</u>: W.B. Heginbotham, "Robots and Automatic Factories",
<u>British Business</u>, 24 April 1981.

The distribution of robots by different industrial sectors in Japan in 1976-1979 is shown in table 16.

Table 16. Distribution of robots by industrial
sector in Japan (1976-1979) 6/

(Percentage based on current value)

Sector	1976	1977	1978	1979
Automotive industry	30	34	39	38
Electrical machinery	21	23	24	18
Metal-working machinery	5	6	4	3
Other machinery	4	3	2	3
Metal-working	6	3	7	8
Other engineering	3	5	2	4
Iron, steel and non-ferrous metals	8	7	5	6
Chemicals, petroleum, coal, ceramics, etc.	3	1	-	-
Food, textile and other manufacturing	3	2	2	4
Plastic-moulding products	13	10	10	11
Others	2	2	2	3
Exports	2	4	3	2
Total	100	100	100	100
Value (millions of US dollars)	68.7	105.6	133.1	190.4

Source: E.K. Yasaki, "Japanese push robotics", Datamation, July 1981.

Similar figures for 1980 are given in table 17.

____

6/ Figures concerning the distribution of robots in Japan appear in the following text as they were issued in part one of the present study on August 1983 (ECE/ENG.AUT/12). More recent information has been published in many sources, e.g. A.S. Kosmynin, "Market of metalworking equipment and industrial robots in Japan"; in Bulletin of Foreign Trade Information, No. 1, VNIKI Institute, Moscow 1984 (in Russian).

Table 17. Distribution of robots by industrial
sector in Japan in 1980

(Percentage)

Sector	Share
Automotive industry	38 a/
Electrical engineering	18 a/
Manufacturing of plastics	11
Manufacturing of metals	8
Metallurgy	4
Metalworking	3
Precision engineering	3
Textile manufacturing	3
Others	12
Total	100

Source: "Industrial robots in Japan", Eastern Economist, vol. 76 (1981), No. 20.

a/ According to Kanje Yonemoto (JIRA), the share of robots decreased from 38 per cent in 1979 to 29 per cent in 1980 in the automotive industry and increased from 18 per cent in 1979 to 36 per cent in 1980 in electronics and electrical machinery.

In a comparative study carried out by the Swedish Association of Mechanical and Electrical Industries in 1976 and 1977, information was obtained concerning inventories of total stock of NC-machines and industrial robots in a sample of 26 firms. The sample gave a fair representation of product variety and firm size of the industry and included heavy users as well as establishments with only a few machines installed. Data were collected from each firm on the two latest installations of NC-machines and robots. The reason for this was partly to introduce a random factor into the sampling and partly the belief that later installations would reflect experience from previous ones, thereby representing a higher degree of techno-economic accuracy. The data were collected through a questionnaire combined with interviews with the technical staffs. The study comprised of 47 NC-machines and 43 robots (see tables 18-23).

The production volume of Swedish firms examined was as follows:

Table 18. Production volume of selected Swedish firms

Units per year	Share of observations (percentage)	
	NC-machines	Robots
1 - 100	15	-
101 - 1 000	17	-
1 001 - 10 000	21	7
10 001 - 100 000	43	58
100 001 - 1 000 000	4	28
1 000 001 - 10 000 000	-	7
Total	100	100

Source: H. Selg and J. Carlsson, "Trends in the development of numerically controlled machine tools and industrial robots in Sweden", Computers and Electronics Commission, Department of Industry, 1980.

The above figures show significant differences between the production volumes of NC-machines and of robots. Compared to NC-machines, robot applications were more related to high-volume goods. About half the NC-machines in the sample were installed in manufacturing where production volume was less than 10,000 per year, while this was the case in only 7 per cent of robots.

Concerning the operation cycle (minutes per unit), the following figures were obtained:

Table 19. Distribution by operation-cycle duration

(Percentage share of observations)

Minutes per unit	0.1 - 1	- 5	- 50	51 -	Total percentage
NC-machines	9	19	49	23	100
Robots	40	47	13	-	100

Source: As for table 18.

There was a considerable difference between NC-machines and robots in regard to the length of the operation cycle. The robots were generally used in very short cycles. In 40 per cent of the observations, cycle length was less than one minute and in almost 90 per cent less than five minutes. A reverse tendency was noted for NC-machines, where only one quarter were used in operation cycles of less than five minutes.

A similar divergency between NC-machines and robots was observed when part variation was studied. While 65 per cent of the NC-machines were used in production where the manufactured goods appeared in more than 10 models, this was the case in only 25 per cent of the robots, most of them paint-spraying robots. Sixty per cent of the robots were found to have between one and five variants, the majority having only one variant:

Table 20. Distribution by number of variants

(Percentage share of observations)

Number of variants	1 - 5	6 - 10	11 - 50	51 -	Total percentage
NC-machines	15	21	26	38	100
Robots	58	16	21	5	100

Source: As for table 18.

The following table shows the frequency of reprogramming.

Table 21. Distribution by number of reprogrammings

(Percentage share of observations)

Number of reprogrammings per week	- 1	2 - 10	11 -	Total percentage
NC-machines	9	70	21	100
Robots	51	49	-	100

Source: As for table 18.

In 70 per cent of the observations the NC-machines were reprogrammed between two and ten times per week. In 21 per cent reprogramming was even more frequent. Half the robots were reprogrammed once a week or less (including no reprogramming at all).

Concerning the batch size, the following figures were obtained:

Table 22. Observations of batch sizes

Number of units	Share of observations (percentage)	
	NC-machines	Robots
1 - 10	23	-
11 - 100	30	5
101 - 1 000	40	28
1 001 - 10 000	7	42
> 10 000	-	25
Total	100	100

Source: As for table 18.

As pointed out above, robots were used in manufacturing with comparatively large batch sizes. In 67 per cent of the observations the batch size amounted to more

than 1,000 units, and in 25 per cent it exceeded 10,000 units. In 50 per cent of
the NC-machine cases, the batch size was less than 100 units, and in 25 per cent
less than 10.

To sum up, a considerable difference in application between NC-machines and
robots was observed. The role of NC-machines in flexible manufacturing was far
more important than that of robots. The following table presents an
approximation of the main tendencies in applications of NC-machines and robots.

Table 23. Summary of tables 18 to 22

Indicator	NC-machines	Robots
Production volume (units/year)	$<$ 10 000	$>$ 10 000
Operation cycle (min/unit)	$>$ 5	$<$ 5
Part variation	$>$ 10	1-5
Reprogramming frequency (times/week)	2-10	-1
Batch size	$<$ 100	$>$ 1 000

Source: As for table 13.

Thus, the production volume related to the average NC-machine was found to be
less than 10,000 units per year. The average operation cycle exceeded five
minutes, and the part manufactured appeared in a relatively large – more than
10 – number of variants. The machine was reprogrammed 2-10 times a week and
consequently the average batch size was low – less than 100 units. In this
sample, the average robot was used where production volume exceeded 10,000 units
per year. The operation cycle was relatively short – less than five minutes.
The part appeared in a few variants, or no variants at all. Reprogramming
occurred no more than once a week and in many cases the robot was not
reprogrammed at all. The average batch size amounted to more than 1,000 units.
In the sample, the flexibility of robots was in most cases exploited to only a
minor degree, or not made use of at all, the robot merely serving as fixed
automation equipment.

The interrelation between the growth of NC-machine tools and of robots (with
the addition of flexible manufacturing systems) in the United States and Japan
(and in France for NC-machine tools only) can be seen in table 24.

Table 24. Trends in the use of NC-machine tools, robots and flexible manufacturing systems in selected countries (1973-1990)

Year	Production of NC-machine tools (units)			Robot installations (millions of US dollars)		Market forecast for flexible manufacturing systems (millions of US dollars)	
	United States	Japan	France	United States	Japan	United States	Japan
1973	2 865	2 765					
1974	4 210	3 040	535				
1975	4 316	2 182	612				
1976	3 856	3 286	576	18			
1977	4 482	5 436	574	27	47		
1978	5 688	7 336	867	31	67		
1979	7 174	13 514	1 068	65	102		
1980				92	266	175	200
1985a/						280	1 050
1990a/						1 200	3 250

Source: D.Z. Beckler, "The Electronic Revolution in the Workplace", OECD Observer, No. 115, March 1982.

a/ Estimate.

A 1979 forecast of the growth of programmable and adaptive robot-system applications in the Federal Republic of Germany, shown in Table 25, includes also an indication of its impact on the evolution of employment.

Table 25. Probable evolution of robot installations and workplaces in the Federal Republic of Germany (1980-1990)

(Thousands of units)

		1980a/	1985a/	1990a/
Robot installations	Final assembly of vehicles	5-6	20-25	35-45
	Manufacture of parts	1-2	5-7	10-15
	Assembly of small parts	-	5-10	20-30
	Machine tools	1-2	5-10	20-30
	Others	1-2	2-4	5-10
	Potential robot installations	8-12	37-56	90-130
Workplaces	Potential workplace losses	16-30	40-160	180-490
	Forecast of new workplaces			
	- Robot producers	2-3	5-7	15-20
	- Control and service personnel	1-1.5	4-6	10-15
	- Total	3-4.5	9-13	25-35

Source: Machintosh Consultants Co. Ltd., Luton 1979.

a/ Estimate.

However, the actual figures showed that the above-mentioned forecast was too optimistic - the total population of programmable robots was only 2,300 in 1981, 3,500 in 1982 and 4,800 in 1983 (see Table 30).

Table 26 shows how the forecast of the market potential of industrial robots in the Federal Republic of Germany (1985 and 1990) points to the tendency towards future concentration of robots in main sectors, such as automotive industry (motor vehicles), electrical engineering, mechanical engineering and earth-moving machinery.

<u>Table 26</u>. <u>Market potential for the use of industrial robots
in the Federal Republic of Germany (1985-1990)</u>
(Units)

Application field	Tool handled by robot		Workpiece gripped by robot		Total	
	1985	1990	1985	1990	1985	1990
Plastics	50	395	925	2 625	975	3 020
Glass			40	100	40	100
Ceramics	55	55	13	150	68	205
Foundries	60	135	460	670	520	805
Steel shaping			350	900	350	900
Mechanical engineering	380	1 500	1 350	3 800	1 730	5 300
Motor vehicles	3 585	7 570	1 360	4 900	4 945	12 470
Earth-moving machinery	479	1 200	950	2 750	1 429	3 950
Precision and capital engineering	50	180	30	110	80	290
Electrical engineering	700	3 700	1 640	5 700	2 340	9 400
Woodworking	25	310	95	875	120	1 185
Pulp and paper			23	265	23	265
Total	5 384	15 045	7 236	22 845	12 620	37 890

<u>Source</u>: <u>Microelectronics, Robotics and Jobs</u>, ICCP (Information, Computer, Communication, Policy) Series No. 7, OECD, Paris 1982.

The market potential of industrial robots in France in 1990 (as compared with the distribution by main sectors in 1980) is depicted in Table 27.

Table 27. Market potential for the use of industrial robots in France (1990)

(Percentage)

Application field	Share of robots installed		
	1980	1990[a]	Change
Food industry	b/	2	+2
Coal, petrol, chemicals	b/		
Metallurgy	b/	28	+10
Metal transformation	10		
Mechanical engineering	8		
Instrumentation	b/	19	+8
Electrical industry	6		
Electronics industry	5		
Shipbuilding	b/	38	-23
Automotive industry	58		
Aeronautical and spatial	1		
Other transport equipment	2		
Hand tools	1	12	+6
Textile industry	b/		
Ceramics and materials	5		
Wood processing	b/		
Paper and printing	b/		
Plastics	4	1	-3
Other	b/		
Total	100	100	0

Source: P. Piernaz, "Le nouvel échiquier de la robotique". L'Usine nouvelle, 9 June 1983.

a/ Estimate.

b/ Negligible.

Some interesting facts concerning the distribution of robots by application areas in western Europe and the socio-techno-economic impact of robotics had also been described in a publication issued by the Commission of the European Communities. 7/

7/ Robotics - current events in Federal Republic of Germany, France, Italy, Ireland, Scandinavia and the United Kingdom: Social Change and Technology in Europe. Information Bulletin No. 10, Commission of the European Communities, Brussels, November-December 1982.

V. WORLD-WIDE DIFFUSION OF ROBOTS

The industrial-robot market is growing steadily all over the world; robot producers include both large manufacturing companies specializing mainly in other fields (automotive industry, electrical industries, general engineering) and small manufacturers dealing especially with sophisticated automation equipment. The present chapter gives some estimates of the current world robot population and forecasts of its growth over the next decade. These figures concern production and use by countries and are based on generally available information. The level of international collaboration in the field is very high owing to licensing and intergovernmental agreements and the growing activities of transnational companies in the sector. The total number of robots in service in 1981 has been estimated at 30,000 units, in 1982 at 39,000 units and in 1983 at 50,000 units excluding manual manipulators and equipment with a fixed, unchangeable sequence of operations (see table 32). With the inclusion of non-programmable manipulators, the total number of units may be four to five times larger. If the present trend continues, i.e. if the total number of units is doubled every two years, around the year 2000 more than 10 million robots could be expected to be in use. 1/

According to other sources, some 1 million robots can be expected to be in use by 1990. A more precise forecast, published by OECD in 1983, is presented in table 28.

Table 28. **Estimate of robot population in selected countries - OECD (1985 and 1990)**

(Units)

Country	1981	1985[a]	1990[a]	Annual growth rate (percentage)	
				1981-85	1985-90
Japan	9 500	27 000	67 000	30	20
United States	4 500	15 000	56 000	35	30
Sweden	1 700	4 100	8 300	25	15
Germany, Federal Republic of	2 300	8 800	27 000	40	25
United Kingdom	713	2 700	10 000	40	30
France	790	2 100	6 500	28	25

Source: Industrial Robots: Their Role in Manufacturing Industry, OECD, Paris 1983 (Chapter IV) - OECD estimate based on data from British Robot Association and Diebold France.

a/ Estimate

____

1/ La Suisse, special supplement, Geneva, 24 June 1982.

A less optimistic forecast of the growth of the robot population in selected countries was issued by UNIDO, as shown in table 29.

Table 29. Estimate of robot population in selected countries - UNIDO (1985 and 1990)

(Units)

Country	1985	1990
Japan	16 000	29 000
United States	7 715	31 350
Germany, Federal Republic of	5 000	12 000
Switzerland	600	5 000
Sweden	2 300	5 000
Norway	1 000	2 000
United Kingdom	3 000	21 500

Source: J. Bessant, "Technolocy and Market Trends in the Production and Application of Information Technology. UNIDO Microelectronics Monitor, No. 8 Supplement, Vienna, December 1983 (based on RIA).

The world robot population reached in 1981-1983 (excluding the centrally-planned economy countries) is shown in table 30.

Table 30. Robot population in selected countries (1981-1983)

(Units)

	1981	1982	1983
Japan	10 000	13 000	16 500
United States	5 000	6 250	8 000
Germany, Federal Republic of	2 300	3 500	4 800
Sweden	1 700	1 300 a/	1 850
United Kingdom	713	1 152	1 753
France	600	950	2 150
Italy	450	700 b/	1 800
Other countries	1 500	1 550	2 000 c/
Total	22 263	28 402	38 853

Sources: - Robot News International, Vol. 2, No. 8, March 1982 (1981 data).

- British Robot Association, Robotfacts, December 1982 (1982 data).

- The Industrial Robot, Vol. 11, No. 1, March 1984 (1983 data).

a/ According to the data received by the ECE secretariat from the Swedish Computers and Electronics Commission, the total population of robots in Sweden was as follows: 1,250 units in September 1981, 1,600-1,700 units by the end of 1982, with some 2,300 robots forecast for 1984.

b/ According to the data supplied by the Italian Society for Industrial Robots (SIRI), there were 1,200 servo-programmable robots installed in Italy in 1982 (i.e. robots that can be programmed from a computer console or by means of a jogstick or similar device, with or without self-adaptive capabilities).

c/ ECE secretariat estimate.

Another source 2/ assessed the total population for 1981, including the USSR, at 28,000 to 32,000 units, with an annual growth of approximately 30 per cent. Slightly lower figures on the world robot population in 1981 (including Czechoslovakia and Poland and excluding the USSR) have also been published. 3/ The total number of units is assessed by the second source at between 22,000 and 23,000. Concerning the distribution of robots by applications, the same source estimates that in Japan the highest share of robots is used for machine loading, followed by assembly and welding; in the United States the order of usage is: welding, machine loading, foundries; in the Federal Republic of Germany: welding, machine loading, painting and finishing operations.

Another estimate of the robot population in 1981 (including that of the USSR and the socialist countries of eastern Europe) together with projected robot usage in 1985 was published in 1982 (table 31).

Table 31. Estimate of robot population in selected areas
and countries (1981-1985)

Countries	Percentage of 1981 world population a/	Equivalent in number of units	Forecast for 1985 in number of units b/
Japan	57.5	14 232	65 000
United States	16.6	4 109	50 000
Western Europe	11.4	2 822	30 000
USSR	12.1	2 995	
Eastern Europe (excluding USSR)	2.4	594	
Total	100.0	24 752	

Source: Manufacturing Automation, Financial Times Survey, 16 July 1982.

a/ Robot Institute of America (RIA).

b/ Predicasts.

An estimate of the world-wide robot population by main regions is given in Table 32, which shows that the total number of robots (excluding simple manipulators) reached some 50,000 units in 1983. It is also estimated on the basis of current trends that the total number of units will probably double every two years and, as from 1986 to 1990, every three or four years. The compound annual growth rate (number of units) for western Europe for the period 1981-1986 was estimated at 37 per cent, 4/ leading application areas being assembly (58.6 per cent), machine tools (51.6 per cent), metal processing (37.1 per cent) and arc welding (36.9 per cent); annual growth rate in the currently most robotized area of spot welding (namely in the automotive industry) will decrease to 6.1 per cent.

2/ P. Vicentini, "I robot nell'industria dell'automobile: una breve panoramica". Notiziario Tecnico AMMA, No. 10, Torino 1981 (Speciale robotica).

3/ "The next step in factory automation", The Economist, 19 December 1981.

4/ Creative Strategies International.

Table 32. ECE estimate of world-wide programmable-robot population (1981-1983)

Region	1981		1982		1983	
	Percentage	Units a/	Percentage	Units a/	Percentage	Units a/
Asia (including Japan)	38.3	11 500	34.6	13 500	33.0	16 500
North America	18.3	5 500	17.9	7 000	17.0	8 500
Western Europe	21.7	6 500	23.1	9 000	26.0	13 000
Eastern Europe (including USSR)	20.0	6 000	21.8	8 500	21.0	10 500
Others	1.7	500	2.6	1 000	3.0	1 500
Total	100.0	30 000	100.0	39 000	100.0	50 000

Source: ECE secretariat estimate.

a/ Rough estimate.

Slightly modified figures on total world population of industrial robots have been recently published in Austria, as shown in table 33.

Table 33. Estimated growth of world robot
population (1974-1985)

(Units)

Country/region	1974	1978	1979	1980	1981	1982	1985 a/
Japan	1 500	3 000	4 000	5 500	8 500	12 000	35 000
United States	1 200	2 500	3 000	4 500	7 000	9 000	28 000
Western Europe	800	2 000	2 650	4 000	6 000	9 000	25 000
Eastern Europe (incl. USSR)	...	...	...	...	6 000	8 000	21 000
Others	...	...	...	...	1 500	2 000	4 000
Total	...	...	...	...	24 000	40 000	113 000

Source: H. Weseslindtner: "Industrierroboter; Entwichlung u. Herstellung in Österreich". Technische Universität, Wien 1984.

a/ Rough estimate.

The following assessment of the density of programmable-robot installations was published early in 1979.

Table 34. Relation between robot installation and employment
in selected countries (1979)

Country	Number of installed robots	Number of employees in the metalworking industry (thousands)	Number of robots per thousand employees
Sweden	700	470	1.49 a/
Japan	4 000	3 175	1.26
United States	2 000	8 320	0.24
Italy	350	1 630	0.21
Germany, Federal Republic of	600	3 930	0.15
United Kingdom	120	3 300	0.04
France	60	2 350	0.03

Source: Mackintosh Consultants Co. Ltd., Luton 1979.

a/ The leading position of Sweden was confirmed also for 1982; see "Un robot sur cinq dans le monde sera bientôt suédois" (Soon one robot in five will be Swedish), Journal de Genève, 26 July 1982.

However, although these figures are already outdated, they give some indication of the correlation between robot installation and employment in various countries.

According to an OECD study, 5/ the number of robots for every thousand employees
in the manufacturing industries in 1981 increased to 3.0 in Sweden, 1.3 in Japan,
0.5 in the Federal Republic of Germany and 0.4 in the United States.

Table 35 shows slightly changed order of countries in 1984 (based on total
number of workers in industry).

Table 35. Relation between robot installation and employment
in industry in selected countries (1982)

Country	Number of installed robots	Industrial workforce (thousands)	Density robots per 10,000 workers
Sweden	1 300	1 352	9.6
Japan	13 000	19 556	6.6
Germany, Federal Republic of	3 500	11 334	3.1
Belgium	361	1 322	2.7
United Kingdom	1 152	5 272	2.2
United States	6 250	29 774	2.1
France	950	7 574	1.25
Italy	700	7 787	0.9

Source: Industrial Robot, Vol. 11, No. 1, March 1984.

A Japanese source 6/ assesses the growth rate in robot use at 50 per cent
per annum for the next five years, and at 10 per cent thereafter. A less
"conservative" prediction based on surveys of United States robot users,
manufacturers and research institutes is given in a report. 7/ In comparison
with a robot population of approximately 5,000 in 1981, in 10 years' time
United States industry will have increased the total number of robots in use by
50 times, to reach 250,000 (projected annual sales 60,000 to 80,000 robots).
Total sales of robots increased from $US 100 million in 1980 to $155 million
in 1981 and had been expected to increase by 35 per cent a year, the average
price of each system being approximately $70,000 in 1981. 8/ In fact, it is
expected that the robot market in the United States will grow by about
20 per cent in 1984, attaining approximately $265 million. Between 1985 and
1988, some 25 per cent annual growth is expected. 9/ According to one
source, 10/ today's installed base in the United States (1984) could be in the
range of nearly 20,000 robots, and could possibly reach 100,000 by the year
1990.

5/ "The impact of industrial robots on the manufacturing industries of
member countries", draft of OECD study, Paris, October 1982.

6/ "Advance of robotic army set to move deeper into industry" (report on
the 11th International Symposium on Industrial Robots held in Tokyo).
The Engineer, 29 October 1981.

7/ "Industrial Robots - A Summary and Forecast for Manufacturing Managers".
Tech. Trans Corp., Washington 1981.

8/ "US Firms Begin to Accept Robots". The Engineer, 11 March 1982.

9/ United States Industrial Outlook 1984, chapter 20, Department of
Commerce, Washington.

10/ J. Teresko, "The robots are ready - but where are the factories of the
future?". Industry Week, 29 October 1984.

By almost any definition of a robot, Japan's robot builders certainly produce more than the rest of the world 11/ but because simple material-handling devices are probably included, the estimates of installed robots vary between 6,000 and 65,000 units.

The trends in the production of industrial robots in Japan are shown in table 36.

Table 36. Evolution of robot production in Japan (1968-1990)

Year	Annual production (number of units)	Total number of units installed in Japan a/	Annual value (millions of yen)
1968 b/	200	200	400
1969 b/	400	600	1 500
1970	1 700	2 300	4 900
1971	1 300	3 600	4 300
1972	1 700	5 300	6 100
1973	2 500	7 800	9 300
1974	4 200	12 000	11 400
1975	4 400	16 400	11 100
1976	7 200	23 600	14 100
1977	8 600	32 200	21 600
1978	10 100	42 300	27 300
1979	14 500	56 800	42 400
1980 b/			60 000
1985 c/	48 700	d/	290 000
1990 c/	84 200	d/	525 000

Source: Kikai Sinko Promot. Mach. Ind. Jap., vol. 14 (1981) No. 5.

a/ JIRA and Inbucon sources as published in the Financial Times, 16 July 1982.

b/ These figures do not include exports from Japan, which are estimated at under 2 per cent of annual production.

c/ Estimate.

d/ According to RIA 9/ the total population is projected to attain 100,000 units in 1985 and 327,000 units by 1990.

According to another source, 12/ the production of robots in Japan attained a value of $US 210 million in 1979 and nearly $400 million in 1980. These figures include certain manipulators that accounted for 52 per cent of sales in 1979 and 44 per cent in 1980. Excluding these units, the Japan Industrial Robot Association is projecting production at $2,200 million in 1985 and $4,500 million in 1990. Leaving aside the possible evolution of the exchange rate, 13/ it seems that the growth rate for the period 1981-1990 is underestimated.

11/ J. Teresko, "Robots come of age - but is management ready to put them to work?". Industry Week, 25 January 1982.

12/ Business Week, 14 December 1981.

13/ Exchange rate in July 1983: 1 United States dollar = approx. 240 yen.

Tables 37 and 38 give the distribution of robots in Japan by various types.
These figures show a total of 56,800 manipulators and robots in 1979
(manual-control and fixed-sequence devices).

Table 37. Distribution of robots by type in Japan (1978-1990) 14/

(Percentage based on units)

Type of robot	1978	1979	1985 a/	1990 a/
Manual control	16	7	11	9
Fixed sequence	70	74	57	56
Variable sequence	6	8	18	17
Playback robots	5	5	7	8
NC-robots	-	1	2	2
Intelligent robots	3	5	5	8
Total	100	100	100	100

a/ Estimate.

Table 38. Distribution of robots by type in Japan (1978-1990)

(Percentage based on current value)

Type of robot	1978	1979	1985 a/	1990 a/
Manual control	5	5	5	4
Fixed sequence	46	47	29	27
Variable sequence	20	18	26	22
Playback robots	17	17	21	20
NC-robots	1	4	5	4
Intelligent robots	11	9	14	23
Total	100	100	100	100

a/ Estimate.

14/ According to an EEC source, the total population of programmable
robots in Japan in 1980 reached 76,661 units, of which 81 per cent were
manually controlled manipulators and fixed-sequence robots (first two lines
of Tables 37 and 38), i.e. 14,250 programmable robots ("L'industrie
Européenne de la machine-outil. Prise de position de la Commission -
Situation et perspectives", EEC document SEC(83)151 final, Brussels,
8 February 1983).

Concerning the annual production of robots in Japan from 1974 to 1979 by various types (see table 39), one may note a tendency towards a more rapid growth in the production of more sophisticated types.

Table 39. Annual growth rate of robot production
in Japan (1974-1979)

(Units)

Type of robot	1974	1975	1976	1977	1978	1979
Manual manipulators	713	772	697	1 127	1 576	1 051
Fixed sequence	3 287	3 297	6 199	6 494	7 066	10 721
Variable sequence	-	-	-	425	652	1 224
Playback robots	165	137	183	357	506	662
NC-robots	1	0	6	11	25	89
Intelligent robots						
Total	4 167	4 418	7 165	8 613	10 100	14 535

Source: Japan Industrial Robot Association.

A recently published forecast regarding the share of robot installations in western Europe gives the figures indicated in table 40.

Table 40. Estimated distribution of robots by type
in western Europe (1981-1986)

(Percentage)

Type of robot installation	1981	1983*	1986*
Pick-and-place robots	55	33	22
Sophisticated robots	37	45	42
Sensor-based robots	-	2	11
Assembly robots (with or without sensors)	8	20	25
Total	100	100	100

Source: Financial Times, 16 July 1982.

According to the British Robot Association's figures, the number of robot systems installed in the United Kingdom increased from 371 in 1980 to 1,753 in 1983, i.e. an average annual increase of approximately 69 per cent. According to the same source, the origin of robots installed in the United Kingdom in 1983 was 28 per cent domestic, 24 per cent imported from the United States, 16 per cent from Japan and 32 per cent from other European countries. The estimates for 1985 vary between 2,500 and 5,000 robots 15/ with many suppliers leaning towards the lower figure. The automotive industry (37 per cent share of the market) and the metal goods industry (17 per cent) were the two largest users in 1981. 16/

15/ M. Wildish, "Robots: a fight for survival". The Engineer, 6 May 1982.

16/ "Robotics in the UK". The Industrial Robot, March 1981.

The growth of industrial-robot installations in Sweden may be seen in table 41. The number of installed robots is expected to increase rapidly; however, the annual growth rate will decrease from 35-39 per cent (1971-1979) to 17-26 per cent (1980-1990). Among the reasons for this lower rate are the following:

- Growth is calculated from a much higher level of robot population;

- There are other automation alternatives besides general-purpose robots. No doubt there are great potential advantages in automating, for instance, assembly and inspection operations with sensor-equipped robots. However, the installation of computer-aided special-purpose assembly machines, the redesign of products and components or the use of new materials may in many situations be the most advantageous ways of improving the efficiency of assembly operations.

Table 41. Robot population in Sweden (1970-1990)

Year	Number of robots	Average annual growth (Percentage)
1970	55	35
1973	135	
1977	490	38
1979	940	39
1984 a/	2 300	20
1990 a/	6 000-9 000	17-26

Source: The promotion of robotics and CAD/CAM in Sweden. Industri-Departement, DSI 1981: 26.

a/ Estimate.

The total robot population in the Netherlands increased from 58 robots in 1981 to 77 in 1982. More than 60 per cent of robots are used for welding and spray-painting operations. 17/ All robots in the Netherlands are imported, with the exception of two assembly robots produced and implemented in 1982.

In France, considerable emphasis is put on the production and use of robots. Main robot applications are in assembly - more than 50 per cent including simple pick-and-place operations - and in loading and unloading of various machines - more than 30 per cent. 18/ The principal users are in the

---

17/ "Flexibele automatisering in Nederland", edited by G. Laurentius, H. Timmerman and A.A.M. Vermeulen, Toekomstbeeld der Techniek, No. 34, Delftse Universitaire Pers 1982 (in Dutch).

18/ Industrial Robots International, Vol. 3, No. 7, 12 April 1982.

automotive industry, namely Renault 19/ (around 60 per cent of total), with
the rest mainly in the electronics industry and the mechanical-assembly
industries. An interesting survey on the state of the art in this field was
published in March 1983. 20/

The robot population in the Union of Soviet Socialist Republics, including
simplified robots, was estimated at 6,000 robots in 1980. 21/ It is believed
that, in the metalworking industries alone, 120,000 robots will be needed in the
next decade, with a view to replacing approximately 350,000 workers with annual
cost savings of about 1,000 million roubles. 22/ If we take into account the
needs of other sectors - steel industry, mining, light industries including the
food industry, building industries, transport, medical services, general
services etc. - the above-mentioned estimate will increase to 350-400 thousand
robots, replacing nearly 1 million workers, with annual cost savings of
3,000 million roubles.

Within the Ministry for the Electrotechnical Industry of the USSR alone,
20 different types of robots and manipulators have been produced since 1973. 23/
They are used mainly in metalworking, forming, pressure casting and assembly
processes. Another source 24/ states that during the first quarter of 1982 a
production of 1,125 automatic programme control manipulators (industrial
robots) was recorded in the USSR: an increase of approximately 90 per cent.
Growth of this order may lead to a 50 per cent USSR share of the world stock
of industrial robots by 1985. According to the Annual Review of Engineering
Industries and Automation, 25/ more than 2,700 automatic industrial manipulators
with digital programme control were produced in the USSR between 1976 and 1980.
The same figure is mentioned in the EKO Journal. 26/

––––––––––

19/ "France makes Renault its model", Business Week, 31 May 1982.

20/ "La Science des robots", Science et Vie, Excelsior Publications,
Paris, March 1982.

21/ E. Jurevitch, "Strategy of robototechnics", Socialisticheskaja
industrija, 14 October 1980.

22/ The total robot population in CMEA countries by 1990 is estimated
at 200,000 units; see O. Rybakov and B. Vjazorov, "Perspectives of robot
applications", Ekonomicheskaja gazeta, No. 26, June 1982.

23/ V. Gavrilov, "Robots and Economy", Socialisticheskaja industrija,
30 September 1981.

24/ Quarterly Economic Review of USSR (second quarter 1982), EIU,
London 1982.

25/ Annual Review of Engineering Industries and Automation 1980
(United Nations publication Sales No. 77.II.E.18).

26/ EKO Journal, Novosibirsk, No. 2, February 1982 (in Russian).

In the Union of Soviet Socialist Republics, the percentage of simplified robots (used mainly for simple manipulations, with a maximum of four articulations and less than 100 programming instructions) will decrease in 1980-1985, while the percentage of multifunctional programmable industrial robots (with five or more articulations, based on sophisticated computer control) as well as so-called adaptive robots (with sensory perception) will increase (see Table 42).

Table 42. Distribution of robots by type in the Union of
Soviet Socialist Republics (1980-1985)

(Percentage based on units)

Type of robot	1980	1981	1982	1985 a/
Simplified robots	77	75	70	60
Multifunctional programmable robots	20	21	25	30
Adaptive robots	3	4	5	10
Total	100	100	100	100

Source: Y.G. Kozyrev "Perspectives of Development and Use of Industrial Robots". Submitted by the Government of the USSR.

a/ Estimate.

In the German Democratic Republic, between 9,000 27/ and 15,000 28/ robots are in use today, while for 1981-1985 the production and use of 40-45 thousand industrial robots is planned. 29/ The diversity in these figures results from differences in the definition of robots (manual manipulators and non-programmable fixed-sequence robots are probably included). It is expected that the introduction of industrial robots could reduce work places by 100,000, i.e. one robot or manipulator will replace approximately 2.5 workers. 30/

The annual production of robots in Czechoslovakia reached 285 in 1981, 31/ including 220 simple manipulators. In 1985 the total population will reach 3,000 robots, replacing some 5,500 workers and for 1990 it is estimated at 14,000 robots replacing approximately 25,000 workers. 32/

27/ Quarterly Economic Review of East Germany (first quarter 1982), EIU, London 1982.

28/ P. Verner, Report presented to the session of the SED Central Committee (EE/7061/C/2), 25 June 1982.

29/ Economic Survey of Europe in 1981, UN/ECE, Geneva 1982, page 227.

30/ Scientific organization of work and implementation of robots in the German Democratic Republic. Poduikova Organizace No. 8, Prague 1984.

31/ J. Winter, "Robot '82". Mladá Fronta, 13 March 1982.

32/ A. Kondrashov, Robots: Yesterday, Today, Tomorrow. Socialisticheskaja industrija, 2 November 1982.

Czechoslovakia has a widely based State target programme of robotization for the seventh five-year-plan period (1981-1985) which concentrates on all stages of the development-production-use cycle. It is believed that the process of robotization "will result in higher labour productivity, less energy-intensive production and the elimination of heavy and high-risk labour, etc.". 33/

There are certainly more robots installed in other countries also. In March 1979 it was estimated that 720 robots were in use in Poland, 200 in Norway and 130 in Finland. 34/ Special measures are undertaken to encourage the use of robots in Ireland, 35/ Australia, Israel and other countries. 36/ According to one source, 37/ there are 528 industrial robots - 316 play-back and 212 pick-and-place - currently installed in Australia (1984). In addition, there are some 520 educational robots (450 of them of Australian manufacture: Tasman Turtle) and 13 personal robots (so-called Androbots) in operation. At the 12th International Symposium on Industrial Robots, 38/ interesting papers were presented concerning the development of robotics in Switzerland, China and Romania. Information on recent developments in this field in Buglaria was presented in April 1982 to the conference "Robots in the Automotive Industry". 39/ It is expected that the number of robots installed in Bulgaria will attain 2,000 units by the end of 1985. 40/ The main application fields are metalworking, plastic moulding, paint spraying, welding and assembly.

As mentioned above, the ECE secretariat circulated two questionnaires requesting data on the diffusion of robots in member countries by operational modes and by application fields. The answers obtained are presented in Chapter VI of the present study.

33/ V. Valek, _Czechoslovak Foreign Trade_, Prague 1982.

34/ R.Y. Horiguchi, "Commanding Lead in Robot Race". _International Herald Tribune Survey_, September 1981.

35/ "Ireland Woos the Robotics Companies". _The Engineer_, 20 May 1982.

36/ _Industrial Robots International_, Vol.3, 1982.

37/ M. Kassler, "Australia's Robots and their International Significance". Proceedings of the 14th International Symposium on Industrial Robots, Gothenburg, 2-4 October 1984, co-published by IFS and North-Holland.

38/ Proceedings of the 12th International Symposium on Industrial Robots, Paris, 9-11 June 1982, published by IFS London.

39/ Proceedings of the International Conference "Robots in the Automotive Industry", Birmingham, 20-22 April 1982, published by IFS London.

40/ T. Chakyrov, "Electronics and Robotics in Production in Bulgaria", Economic Co-operation of CMEA Member Countries. Journal CMEA Moscow, No. 1, 1984.

A study prepared by the secretariat of the United Nations Conference on Trade and Development (UNCTAD) also describes the possible impact of robotics. 41/ It suggests that the common requirement for the successful utilization of modern production and design technologies (i.e. NC-machine tools, robots and computer-aided design) is the ability to operate and maintain the sophisticated equipment and to use and develop associated software. As noted, "due to unskilled labour saving property of robots (with some exceptions), their diffusion to the developing countries would not seem to accelerate in the near future". 42/

The robotics market is growing steadily and, in western Europe alone, it will probably attain an annual sales figure of $US 2,000 million in 1990 (in 1981 prices). An estimate of robot sales in western Europe for the period 1981-1986 is given in table 43.

Table 43. Estimate of robot sales in western Europe (1981-1986)

Year	Number of units	Annual value of sales (millions of US dollars)
1981	1 700	96.0
1982	2 516	166.0
1983	3 270	224.8
1984	4 251	309.6
1985	6 291	558.6
1986	8 178	763.2
Total for 1981-1986	26 206	2 118.2

Source: Creative Strategies International.

It is not the aim of this study to undertake a review of types and makes of robots. The robotics industry today includes hundreds of producers, some of them highly specialized in this field for many years. What should be noted as a significant tendency (leaving aside new entries, abandonments, mergers etc.) is the entry into the robotics market of such large companies as International Business Machines, specialized in information technology, General Motors, leading automotive producers, and General Electric, dealing with electrical supplies and investments. The IBM 7535 manufacturing system (in collaboration with Japan) and the RS 1 robot system 43/ are understood to represent an important stage in IBM's over-all CAD/CAM strategy. The 7535 system is based on the IBM-developed language AML (A Manufacturing Language). This language has the ability to adapt to its working environment through various optical and tactile sensors. Application-development programmes within the AML are available also for the RS 1 system. Most important for the near future seems to

---

41/ Problems and Issues Concerning the Transfer, Application and Development of Technology in the Capital Goods and Industrial Machinery Sector: The Impact of Electronics Technology on the Capital Goods and Industrial Machinery Sector: Implications for Developing Countries (United Nations Conference on Trade and Development, document TD/B/C.6/AC.7/3), pp. 17-24.

42/ See UNCTAD, document TD/B/C.6/AC.7/L.1.

43/ "I (BM) Robot". Datamation, April 1982.

be the company's ability to utilize a wide software-development background and experience, in view of the fact that the share of software costs is expected to increase from the present 50 per cent to nearly 90 per cent of over-all system costs. The General Electric Company (in co-operation with companies in Japan, Italy and the Federal Republic of Germany) has become one of the most important robot manufacturers. 44/ The range of General Electric robots consists of 12 models, which can handle loads of from 15 to 100 kg; an example is the model GP 132, a seven-axis robot with electric servo-motors. 45/ General Electric has also shown its new vision-memory system Optomate I, made up of a microprocessor-controlled buffer memory and an analog/digital charge injection-device camera. It has a variety of uses including industrial monitoring and robotic inspection systems. The General Motors Corporation already has 1,400 working robots in its manufacturing plants throughout the world and projections are made to have 20,000 robots installed by 1990. 46/ It is obvious that these high-volume demands may be better fulfilled by the use of General Motors' own robots (special facilities, low costs, maintenance etc.). A similar tendency may be seen in the ECE region, 47/ e.g. Comau-Fiat and Olivetti in Italy, Renault in France, Volkswagen and Siemens in the Federal Republic of Germany, ASEA and Electrolux in Sweden.

The same situation prevails in Japan: among 122 robot producers listed in a survey 48/ there are 11 - Kawasaki Heavy Industries, Hitachi, Yasukawa Electric, Mitsubishi Heavy Industries, Fuji Electric, Kobo Steel, Fujitsu-Fanuc, Sankyo Seiki, Nitto Seiko, Shinmeiwa Industry and Star Seiki - which together produce approximately 4,200 programmable robots (1981 target) valued at some $US 130 million (the average price of one robot being $US 31,000). On the other hand, one should not overestimate the future growth rate: according to another Japanese source, 49/ "for the first time since its birth, the Japanese robot industry in 1982, experienced a surprising slowdown in its annual growth production ... only 20 per cent up from the preceding year, compared with 55 per cent in 1979, 85 per cent in 1980 and more than 50 per cent in 1981".

A more detailed review of the world robot industry may be found in Chapter VII of the present study.

____

44/ "GE is about to take a big step in robotics". Business Week, 8 March 1982.

45/ Industrial Robots International, Technical Insights Inc., vol. 3, No. 6, 22 March 1982.

46/ Ibid.

47/ J. Le Quément, "The Social and Economic Stakes in International Competition in the Robotics Industry". Proceedings of the 12th International Symposium on Industrial Robots, Paris, 9-11 June 1982, published by IFS, London.

48/ Robots in the Japanese economy (Facts about robots and their significance), edited by K. Sadamoto. Survey Japan, Tokyo 1981.

49/ M. Shimogama, "Japanese Industrial Robot Industry Now Facing Change". The Japan Economic Review, 15 June 1983.

VI. DIFFUSION OF ROBOTS IN THE ECE REGION

This chapter is based on national contributions received in answer to two questionnaires on the distribution, at the end of 1982, of programmable robots and robots with sensory perception.[1] It contains a review of data and information received from Austria, Belgium, Canada, the Federal Republic of Germany, Hungary, Italy, Ireland, the Netherlands, Portugal, Sweden, [2] the United Kingdom and the United States. A number of other countries indicated that they were not in a position to provide the required data.

Additional information submitted by the secretariat of CMEA and by the Union of Soviet Socialist Republics, i.e.

- Information on multilateral co-operation of the CMEA member countries in the field of robotics

- Perspectives of development and use of industrial robots in the USSR

is included in annexes VII and VIII, respectively.

AUSTRIA

The total population of programmable robots and robots with sensory perception in Austria by the end of 1982 is shown in tables 44 and 45.

---

[1] Annexes I and II

[2] The complete text of the Swedish contribution may be found in annex VI

Table 44. Distribution of robots by operational mode in Austria (1982 and 1983)

(Units)

Operational mode	Number of units	
	1982	1983
Handling and transport operations	18	19
Loading and unloading of machines	15	17
Workpiece gripped by robot		
Subtotal	33	36
Joining, including welding	21	23
Surface treatment, including paint spraying	4	5
Other metalworking	3	3
Others (tool handled)	2	2
Tool handled by robot		
Subtotal	30	33
Assembly operations	7	8
Total	70	77

Table 45. Distribution of robots by application field in Austria (1982 and 1983)
(Units)

Application field	Number of units	
	1982	1983
Motor vehicles	12	13
Electrical and electronic engineering	28	31
Other engineering	21	23
Engineering		
Subtotal	61	67
Plastic products and other chemicals	6	6
Textile and food industry	3	4
Industrial, excluding engineering		
Subtotal	9	10
Total	70	77

BELGIUM

The population of programmable robots in Belgium reached a total of 361 units by the end of 1982 (tables 46 and 47), all of foreign origin.

Table 46. Distribution of robots by operational mode in Belgium (1982)

(Units)

Operational mode	Number of units
Handling and transport operations	10
Loading and unloading of machines	13
Workpiece gripped by robot	
Subtotal	23
Joining, including welding	219
Surface treatment, including paint spraying	22
Other metalworking	3
Other (tool handled)	90
Tool handled by robot	
Subtotal	334
Assembly operations	4
Total	361

Table 47. Distribution of robots by application field in Belgium (1982)

(Units)

Application field	Number of units
Motor vehicles	226
Electrical and electronic engineering	7
Non-electrical engineering	56
Engineering	
Subtotal	289
Ferrous and non-ferrous metals	12
Plastic products	14
Other industrial, excluding services	46
Industrial, excluding engineering	
Subtotal	72
Total	361

CANADA

The Canadian authorities transmitted to the secretariat a list of robot installations, as at March 1983, by user, robot type, application field and year of installation (see table 48). According to the data received, there were some 126 units installed. Canada informed the secretariat that a feasibility study on the measurement of automated manufacturing was currently being undertaken, and that it should therefore be possible for them to respond to the ECE questionnaire in the near future.

Table 48. Robot installations in Canada (March 1983)

User and plant location	Robot make and model	Number of units	Application key	Year of installation
Ahoy Industries Corp., Richmond, B.C.	Cincinnati Milacron T3	1	B	1982
Atomic Energy of Canada Ltd., Chalk River, Ont.	Cincinnati Milacron T3	1	A	1982
CN Rail, Moncton, N.B.	Unimate 40068	1	A	1979
Centennial College of Applied Arts) Technology, Scarborough, Ont.	Cincinnati Milacron T3 Prab 4200	1 1	E E	1982 1982
Century Machine Co. Ltd., Calgary, Alta.	Fanuc M Model 0	1		1983
Chrysler Canada Ltd., Windsor, Ont.	Prab FB	18	B	1983
Cominco Ltd., Trail, B.C.	Cincinnati Milacron T3 Cincinnati Milacron HT3	1 1	B B	1980
Duplate Canada Inc., Oshawa, Ont.	Prab FA	1	A	1982
Eaton Yale Ltd., Walleceburg, Ont.	Prab 4200	1	A	1980
Eaton Yale Ltd., Suspension Div., Chatham, Ont.	Prab 4200	1	A	1981
Fabco Metals, Windsor, Ont.	Schrader Bellows	1	A	1983
Fisher Gauge Ltd., Peterborough, Ont.	Schrader Bellows Motionmate	1 4	A	1983

Table 48 (cont'd)

User and plant location	Robot make and model	Number of units	Application key	Year of installation
Ford Essex Casting, Windsor, Ont.	Prab 4200	1	A	1982
Ford Motor Co. of Canada Ltd., Oakville, Ont.	Prab FA	4	B	1971
Freight Master Canada Ltd., St. Stephen, N.B.	Cincinnati Milacron T3	1	B	1981
General Motors of Canada Ltd., Oshawa, Ont.	Cincinnati Milacron T3 Cincinnati Milacron HT3 Prab 4200 Prab FA	6 18 4 2	A B C A B C A A	1979 1980 1981 1981
General Motors of Canada Ltd., Ste. Therese, P.Q.	Cincinnati Milacron HT3	1	B	1980
General Motors of Canada Ltd., Windsor, Ont.	Cincinnati Milacron T3 Cincinnati Milacron HT2	14 10	A A	1980/81 1980/81
ITT Almco Division Mississauga, Ont.	Prab 4200	2	A	1983
Imperial Clavite Canada Inc., St. Thomas, Ont.	Prab 4200	3	A	1980/81/82
Inland Steel & Forgings Ltd., Winnipeg, Man.	Schrader Bellows	1	A	1982
International Harvester Canada Ltd., Hamilton, Ont.	Prab FB	6	A	1981
Kelsey-Hayes Canada Ltd., Woodstock, Ont.	Prab 4200	2	A	1979
Kenroc Tools Inc., North Bay, Ont.	Okuma LC-40 18	1	A	1983
McDonnell Douglas Canada Ltd., Toronto, Ont.	Cincinnati Milacron T3	1	B	1981
National Research Council Canada, Computer Graphics, Ottawa, Ont.	Puma 580	1	F	1981

Table 48 (cont'd)

User and plant location	Robot make and model	Number of units	Application key	Year of installation
National Research Council Canada, Industrial Materials Research Institute, Montreal, P.Q.	ASEA IRB-80	1	B	1982
National Research Council Canada, Systems Lab, Ottowa, Ont.	Puma 580	1	F	1981
National Steel Car Ltd., Hamilton, Ont.	Cincinnati Milacron T3	1	B	1980
Niagara College of Applied Arts & Technology, Welland, Ont.	Schrader Bellows Motionmate	1 3	E	1983
Peterson Spring Service Ltd., Toronto, Ont.	Prab 4200	1	A	1983
Ryerson Polytechnical Institute, Toronto,Ont.	Rhino	1	E	1982
Belox Manufacturing Ltd., Toronto, Ont.	Unimate Unimate Apprentice	1 1	B B	1981 1982
Teledyne Canada Metal Products Division, Woodstock, Ont.	Cincinnati Milacron T3	1	B	1982
University of New Brunswick, Dept. of Mechnical Engineering, Fredericton, N.B.	Unimation Puma 600	1	E	1982

Application key

A Parts and material handling

B Tool handling and process applications, including welding, spray-painting, grinding, polishing and de-burring

C Testing and inspection

D Assembly

E Education

F Research

Source: Canadian Machinery and Metalworking, March 1983.

FEDERAL REPUBLIC OF GERMANY

The total robot population in the Federal Republic of Germany at the end
of 1982 is given in table 49.

Table 49. Production, export and import of robots in
Federal Republic of Germany (1982)

(Units and value)

Indicator	Number of units	Value in thousands of DM a/
Production 1982	1 600	260 000
Exports 1982	650	105 625
Imports 1982	400	65 000
Total robot population	3 500	550 000

a/ Exchange rate: 1 DM = $US 0.40 approx.

According to the information provided by the British Robot Association in
United Kingdom, the total population of robots installed in the Federal Republic
of Germany increased from 541 in 1977, 1255 in 1980 and 2300 units in 1981 to
3500 in 1982. Their distribution by operational mode is shown in table 50.

Table 50. Distribution of robots by operational mode in
the Federal Republic of Germany (1982)

(Units)

Operational mode	Number of units
Die-casting	120
Machine-tool servicing	193
Press-tool servicing	70
Forging	52
Other handling	520
Spot welding	1 331
Arc welding	585
Surface coating	397
Grinding, fettling etc.	20
Assembly	122
Education and research	58
Other	32
Total	3 500

Source: IPA, Stuttgart

IRELAND

In Ireland, information on investments in machinery is at present listed
under a global heading: "Purchases of plant, machinery, equipment and vehicles".
No separate data is collected on investments in robots and their use in industry.

ITALY

The information received from Italy is based on a survey undertaken by the
Italian Society for Industrial Robots (SIRI) on servocontrolled robots, i.e.
robots programmable from a computer console or by means of a joystick or similar
device, with or without self-adaptive capabilities. Table 51 shows the
distribution of such robots in Italy at the end of 1982.

<u>Table 51</u>. <u>Distribution of servo-programmable robots by
operational mode in Italy (1982)</u>

(Units)

Operational mode	Number of units
Manipulation, handling etc.	200
Point welding	500
Continuous welding	100
Paint spraying	100
Assembly	130
Measuring	30
Other	140
Total	1 200

The price of robots currently installed in Italy varies between 30 and 150
million lire per unit, the less expensive units being used for simple handling
operations and the most expensive for measurement, assembly and some paint
spraying and welding operations.

HUNGARY

The distribution of programmable robots in Hungary in 1982 is shown in table 52.

Table 52. Distribution of programmable robots by operational mode in Hungary (1982)

(Units)

Operational mode	Number of units
Loading and unloading of machine tools	3
Welding robots	6
Surface treatment, including paint spraying	5
Assembly robots	3
Total	17

NETHERLANDS

Tables 53 and 54 show the distribution of robots by operational mode and application field in the Netherlands at the end of 1982, at which time the total population stood at 77 units.

Table 53. Distribution of robots by operational mode in the Netherlands (1982)

(Units)

Operational mode	Number of units
Loading and unloading of machines	8
Workpiece gripped by robot	
Subtotal	8
Joining, including welding	30
Surface treatment, including paint spraying	21
Other metalworking	8
Other (tool handled)	8
Tool handled by robot	
Subtotal	67
Assembly operations a/	2
Total	77

a/ Assembly robots with sensory perception produced in the Netherlands. All other robots are imported except one surface-treatment robot which is also of domestic origin.

Table 54. Distribution of robots by application field in the Netherlands (1982)

(Units)

Application field	Number of units
Motor vehicles	4
Electrical and electronic engineering	6
Non-electrical engineering	53
Engineering	
Subtotal	63
Plastic products	1
Construction	2
Other industrial, excluding services	5
Industrial, excluding engineering	
Subtotal	8
Research and development	2
Education and training	4
Non-industrial	
Subtotal	6
Total	77

PORTUGAL

The robot population in Portugal totalled 14 units at the end of 1982, including those still in the process of installation (tables 55 and 56).

Table 55. Distribution of robots by operational mode in Portugal (1982)

(Units and value)

Operational mode	Number of units	Value in thousands of escudos a/
Handling and transport operations	6	6 000
Loading and unloading of machines	6	23 000
Other (workpiece gripped)	1	1 000
Workpiece gripped by robot		
Subtotal	13	30 000
Joining including welding	1	8 000
Total	14	38 000

a/ Exchange rate: 1 US dollar = 100 escudos approx.

Table 56. Distribution of robots by application field in Portugal (1982)

(Units and value)

Application field	Number of units	Value in thousands of escudos a/
Electrical and electronic engineering	13	30 000
Non-electrical engineering	1	8 000
Total	14	38 000

a/ Exchange rate: 1 US dollar = 100 escudos approx.

The information received from Portugal also included a list of factors which influence decisions in that country regarding the installation of robots. Those factors are as follows:

- Decreased manufacturing costs;

- Improved productivity;

- Increased competitiveness;

- Replacement of frequently unavailable specialized workers such as welders; and

- Improved product quality.

SWEDEN

As mentioned above, the Swedish contribution has been presented in its entirety as annex VI. The robot population in Sweden totalled 1600-1700 units at the end of 1982, of which 215 had been installed during 1982 (table 57).

Table 57. Production, export and import of robots in Sweden (1982)

(Units and value)

Indicator	Number of units	Value in thousands of Swedish kroner) a/
Production 1982	912	376 000
Exports 1982	805	347 000
Imports 1982	108	36 000
Domestic supply 1982 b/	215	65 000

a/ Exchange rate: 1 US dollar = 7.60 Swedish kroner

b/ Domestic supply = production - exports + imports

UNION OF SOVIET SOCIALIST REPUBLICS

The information received from the Union of Soviet Socialist Republics has been used in Chapter III (table 5), Chapter IV (table 12) and Chapter V (table 42), as well as in figures 4 to 9 and annex VIII.

UNITED KINGDOM

The information received from the United Kingdom is based on data collected by the British Robot Association. The total population of robots installed in the United Kingdom increased from 80 units in 1977, 125 in 1978, 371 in 1980 and 713 in 1981 to 1152 units in 1982. Their distribution by operational mode is shown in table 58.

Table 58. Distribution of robots by operational mode in the United Kingdom (1982)

(Units)

Operational mode	Number of units	
	Installed in 1982	Total installations
Die-casting	3	36
Investment-casting	4	12
Injection moulding	78	167
Machine-tool servicing	55	111
Press-tool servicing	3	27
Palletizing and packaging	14	38
Forging	3	5
Other handling	7	10

Table 58 (cont'd)

Operational mode	Number of units	
	Installed in 1982	Total installations
Spot welding	82	249
Arc welding	70	157
Surface coating	36	124
Grinding, fettling etc.	7	13
Inspection and testing	8	15
Assembly	17	32
Education and research	28	38
Other	24	118
Total	439	1 152

The origin of the robots installed in the United Kingdom is shown in table 59.

Table 59. <u>Distribution of robots by origin in the United Kingdom (1982)</u>

(Units and percentage)

Origin of robots	Installed in 1982		Total installation	
	Number of units	Percentage	Number of units	Percentage
Domestic production	101	23	287	25
Imported from other European countries	164	37	422	37
Imported from the United States	64	15	279	24
Imported from Japan	110	25	164	14
Total	439	100	1 152	100

The distribution of robots installed in the United Kingdom by type of control is shown in table 60. In addition to the 1152 industrial robots, 762 so-called micro non-industrial units have also been implemented.

Table 60. <u>Distribution of industrial robots by type of control in the United Kingdom (1982)</u>

(Units)

Type of control	Number of units
Servocontrolled: point-to-point control	435
Servocontrolled: continuous path control	470
Non-servocontrolled: general purpose robots	114
Non-servocontrolled: Die-casting and moulding machines	133
Total	1 152

UNITED STATES

Table 61 presents the distribution of robots by operational mode installed in the United States.

Table 61. Distribution of robots by operational mode in the United States (1982)

(Units)

Operational mode	Number of units
Handling and transport operations	1 300
Loading and unloading of machines	1 060
Others – workpiece gripped	875
Subtotal	3 235
Joining including welding	2 453
Surface treatment including paint spraying	490
Others – tool handled	50
Subtotal	2 993
Assembly robots	73
Programmable robots total	6 301
Robots with sensory perception	155
Total robot population	6 456

Slightly modified data are presented in table 62.

Table 62. <u>Distribution of robots by type of application
in the United States (1982)</u>

(Units)

Type of application	Number of units
Die-casting	880
Investment-casting	120
Injection moulding, machine-tool and press tool servicing	1 470
Other handling	1 950
Spot welding	1 190
Arc welding	270
Surface coating	290
Grinding, fettling etc.	30
Assembly	50
Total	6 250

<u>Source:</u> Robot Institute of America.

VII. THE WORLD ROBOT INDUSTRY - CHARACTERISTICS, STRUCTURE AND TRENDS

Background

The robot industry is a very young one. The first commercial use of industrial robots - as they are now commonly defined - dates back to the beginning of the 1960s. However, it was not until the middle of the 1970s that output reached a level which warranted the consideration of the industrial robot industry as a separate entity. Before that time, only very few companies were achieving yearly sales in the order of as much as 10 million dollars; orders were received only intermittently from a very thin customer base.

For several reasons, the inception of the industrial-robot industry can be said to date back to the middle of the 1970s:

(a) The results of the research and development work carried out in the period 1960-1975 began to be ready for commercialization;

(b) New microelectronic components became available, especially the microprocessors (developed during the early 1970s) which form the basis of today's powerful and cost-effective control systems;

(c) Time is required for potential users to assess the technology and acquire the necessary knowledge for its use. Not until the middle of the 1970s did the large automotive, electrical and electronics manufacturers (which are still the predominant robot users) commence planning long-term and large investments in robots which, in turn, spurred other firms (e.g. their subcontractors) also to start planning investment in robots;

(d) During the mid-1970s, enterprises started increasingly to seek means of improving productivity in order to compensate for the decline in world market growth (partly due to the "oil crises") and increased world competition (not only from industrialized countries but also from so-called newly industrialized countries).

Thus, at the same time as business strategies were aimed at increasing productivity, robot technology, together with other computer-controlled manufacturing equipment, had attained a degree of perfection which made it commercially ready for wide use in industry.

Since the second half of the 1970s, the robot industry has seen very rapid growth. In terms of value, as well as of units, average yearly growth of robotics has exceeded 30 per cent, attracting the entry into the business of both new and large established firms. According to several forecasts, the high growth rate of robot installations is likely to continue, at least throughout the present decade.1/ 2/ 3/

1/ Competitive position of U.S. producers of robotics in domestic and world markets. USTIC Publication 1475. United States International Trade Commission, Washington D.C., December 1983.

2/ Robots in the Japanese economy. Facts about robots and their significance. Published by Survey Japan, Tokyo 1981.

3/ Robot system business in Japan. Nomura Research Institute, Kamakura City.

Despite high growth rates, the robot industry is still a relatively small industry. Total output in 1982 was only in the order of 3 per cent of that of the machine-tool industry. 4/ However, those aspects which make the robot industry important - one could even say strategic - are the following:

(a) It has a very high growth potential, which may, in the long term, result in its becoming a large industry; and

(b) It has become a symbol of factory automation in general and thus has a considerable indirect effect on the wide diffusion of factory automation equipment.

The robot industry is a high-technology industry which in some respects differs from other comparable high-technology industries such as microelectronics and telecommunications industries. These differences stem mainly from the fact that the development of robotics, to a higher degree than that of the above-mentioned areas, is based on the integration of a multitude of technologies from a variety of industries - machine tools, computers, process control, sensor technology, software and various kinds of application knowledge such as welding, painting, assembly etc.

When considering the world robot industry, one can distinguish common characteristics as well as considerable regional differences with respect to the size of the industry, firm structure and strategies for development, manufacturing and sale. These common characteristics and regional differences are summarized below.

Number of robot manufacturers

There is no exact record of the number of robot-producing firms but it has been estimated that in 1983 there were some 250 in Japan (including 80 which produced robots for their own use only), approximately 50 in the United States and probably the same number each in western and eastern Europe. 1/ 5/ The number of firms on a world-wide basis with external robot sales can thus be estimated to be on the order of 300.

Size structure

Most robot companies (or robot divisions within enterprises) are small. With a total robot turnover of about $150 million in the United States in 1982, the mean turnover of the 50 robot firms was only $3 million. 1/ Amongst those 50 firms, six accounted for approximately 80 per cent of the total output.

Similar concentration rates are found in several other countries. In Japan, in 1982, 14 manufacturers accounted for 65 per cent of the total robot output. 3/ In the Federal Republic of Germany, 90 per cent of total production originated from 10 firms. 1/ In Sweden, a single firm accounted for more than 80 per cent of total output. 4/

4/ Production and use of industrial robots in Sweden in 1982. Computer and Electronics Commission, Ministry of Industry, DsI 1983 :1, Stockholm 1983.

5/ L. Conigliaro, Trends in the robot industry (revisited): where are we now? Proceedings of the 13th International Symposium on Industrial Robots, volume I, Chicago, April 1983.

Origin of robot manufacturers

Before launching out into the manufacture of robots, a firm must ensure the fulfilment of certain preconditions. It must have the capability of bringing together all the various technologies on which robots are based (mechanical, electronic, computer, communication and software engineering). It must also be endowed with extensive know-how in various process applications as well as in the integration and interfacing of the robots in manufacturing systems which might include several different types of machines. The multidisciplinary character of robot manufacturing is a major reason for the fact that enterprises embarking upon robot manufacturing may have a variety of origins (fig. 11).

Concerning the industrial origin of present-day robot manufacturers and the point in time of their decision to enter robot production, some differences may be distinguished between the more important robot-producing countries and regions of the world. These differences are commented briefly below.

The United States. During the 1960s and 1970s, most firms embarking on robot manufacturing had a mechanical engineering background: machine tools, material handling and diversified non-electronic manufacturing (i.e. Unimation, Cincinnati Milacron, Prab and AMF Versatran). It is interesting to note that at that time neither the automobile nor the computer manufacturers were engaged in robot manufacturing for external sale.

The end of the 1970s and the beginning of the 1980s saw large diversified manufacturers, with in-house electronics capabilities, such as Westinghouse and General Electric starting out on the creation of units or divisions for robot manufacturing and marketing. To a large extent these enterprises entered the robot field through the acquisition of independent robot firms and through a network of licensing agreements (concerning both manufacturing and sales) with Japanese and west European robot manufacturers. The structure of the network of licensing agreements is discussed towards the end of the present chapter and depicted in figure 13a-13f.

During the first half of the 1980s, the number of firms entering the robot field increased considerably. Broadly speaking, the new robot manufacturers could be classified in two groups. The first group would consist of large enterprises, notably General Motors and IBM, which had set up either robot divisions or robot subsidiaries. As these enterprises had long been the predominant users of robots, they had the benefit of extensive application knowledge. They had also for several years been carrying out development work in their research laboratories. However, despite their financial and technological strength, in launching out into the manufacture and sale of robots, these companies also had to depend on the acquisition of independent robot firms, and licensing and joint ventures, notably with Japanese robot manufacturers.

The second group of firms, recent entrants to the robot area, would consist of a large number of small start-up firms whose founders frequently had a background in electronics. For various reasons (financial, manpower, competition etc.) these companies have concentrated their activities on certain market segments of the robot market, such as assembly systems, vision systems etc.

Japan. The year of the commencement of robot production in Japan is usually taken to be 1968, when Kawasaki Heavy Industries began to manufacture robots under licence from Unimation Inc. Since then, a large number of firms have entered the robot field. These had their origin both in the machine-tool industry and - unlike most of the American robot firms - in the computer and electronic industry. Most of the major Japanese enterprises in the latter industry, such as Hitachi, Toshiba, Fujitsu, Matsushita, Sankyo Seiki, NEC etc., are also amongst the leading robot producers.

Several of these companies entered the robot field as early as the first half of the 1970s. The above-mentioned companies are large conglomerates producing not only computers and electronics equipment but also telecommunications equipment, machine tools, consumer electronics, household appliances etc. Besides having in-house expertise in mechanical as well as in electronics and software engineering, they are also, together with the automotive industry, the predominant users of robots. They thus had the advantage of possessing considerable application knowledge, which is another important factor in the achievement of a successful robot business. It is usually agreed that the Japanese electronics enterprises started to employ robots earlier that their American counterparts.

In <u>Western Europe</u>, the origin of the robot manufacturers is more evenly distributed over the various industries referred to in figure 11, although there seems to be a bias towards firms having a mechanical-engineering origin. With the exception of ASEA, Olivetti and Siemens, the large electrical and electronics companies in western Europe "have not been at the forefront of technological development during the last decade". <u>6</u>/ However, as distinguished from both the United States and Japan, several west European car manufacturers - Volkswagen, Fiat (Comau) Renault and Volvo - are robot manufacturers as well as users (see table 63).

In his publication "Robots in manufacturing - key to international competitiveness", <u>7</u>/ J. Baranson discusses the differences in the strategies adopted by manufacturers of automated manufacturing equipment and systems (AMES) in the United States, Japan and western Europe. His conclusion is that, whilst American manufacturers have concentrated on equipment and systems with a high degree of sophistication, the Japanese manufacturers have selected the strategy of lower-cost equipment and of close co-operation with the users as well as with manufacturers of AMES components (fig. 12). The strategies of the west European manufacturers, according to J. Baranson, are similar to those of the American as regards co-operation with users and component manufacturers.

<u>A comparison between robot use in the automotive industry and the electrical and electronics industry</u>

As indicated above, Japanese robot manufacturers are much more likely to have their origin in the electrical and electronics industry than their American and west European counterparts. One major reason for this is that the Japanese electrical and electronics industry, which expanded very rapidly during the 1970s, started early to invest in robotics and build up in-house know-how in the field.

<u>6</u>/ G. Junne and R. van Tulder. <u>European multinationals in the robot industry</u>. University of Amsterdam, January 1984. (Draft study sponsored by the Institute for Research and Information on Multinationals - IRM).

<u>7</u>/ J. Baranson, <u>Robots in manufacturing - key to international competitiveness</u>. Lomond Publications Inc., Maryland, United States 1983.

Table 63. Selected major robot manufacturers and their origin

Western Europe	United States	Japan
Automotive:		
Volkswagen (FRG)	General Motors	Toyoda (Toyota)
Fiat Comau (Italy)	Fanuc	
Renault (France)	(GMF) Robotics	
Volvo (Sweden)		
Electrical machinery, electronics, computers; including conglomerates with electronic capabilities:		
ASEA (Sweden)	IBM	Hitachi
livetti (Italy)	Unimation (Westinghouse)	Matsushita Electric
Siemens (FRG)	General Electric	Toshiba
DEA (Italy)		Yaskawa Electric
		NEC
		Fujitsu Fanuc
		Mitsubishi Electric
		Fuji Electric
Mechanical engineering: machine tools, material handling, process technology etc.:		
Kuka (FRG)	Cincinnati Milacron	Kawasaki Heavy Industries
Trallfa (Norway)	Bendix	
	Prab Robotics	

From the data provided in table 64 below, it may be seen that, in 1980, robot shipments to the Japanese electronic equipment industry exceeded shipment made to the automotive industry.

Table 64. Robot shipments to industry in Japan (1976-1980)
(Percentage of total value)

Industry	1976	1977	1978	1979	1980
Auto manufacturing	30	34	39	38	29
Electronic equipment	21	23	24	18	36
All other industries	49	43	37	44	35
Total	100	100	100	100	100

Source: See foot-note 7 above.

Most of the robots applied in the electronic-equipment industry are utilized in assembly and inspection operations. The current trend is towards an increase in the share of total robot shipments being taken by the electronic-equipment industry. 8/ For the sake of comparison between the situation in the United States and Japan, it may be noted that, of total robot sales in 1983 in United States, 6 per cent of shipments were destined for electronic assembly and 6 per cent for all other assembly applications (table 65). 9/ Assembly applications are, however, expected to grow very rapidly and, according to forecasts, are expected to account for 32 per cent of all robot shipments by 1995.

Table 65. Total robot sales in the United States and Shipments
for assembly operations (1983-1995)

Year	Total robot sales (units)	Electronic assembly	All other assembly
		Percentage of total sales	
1983	2 800 (actual)	6	6
1990	10 000 (projected)	12	12
1995	16 000 (projected)	16	16

Source: See foot-note 9 below.

It is estimated that the automotive industry in the United States accounts for from 50 to 60 per cent of the total stock of robots installed in that country. West European countries with large automotive industries show the same pattern as that prevailing in the United States, that is, the automotive industry is still predominant robot user but the electrical and electronics industry is rapidly increasing its share of total robot use. This is illustrated in table 66.

8/ Capital sepnding on robots and business machines rises sharply The Japan Economic Journal, October 26, 1982.

9/ Robotics Today, October 1984.

Table 66.   Industrial robots as a percentage of total robot population
in the Federal Republic of Germany (1980-1994)

(Percentage)

Industry	1980 (approx.)	1985-1987 a/	1990-1994 a/
Mechanical engineering	8	14	14
Motor industry	56	38	34
Earth-moving machinery	8	11	10
Electrical engineering	10	19	24
Other	28	18	18
Total	100	100	100

Source: Micro-electronics, robotics and jobs. OECD, ICCP Series No. 7, Paris 1982.

a/ Estimate.

Investments: reasearch and development, capital equipment and marketing

From the technological point of view, the development of industrial robots is moving towards:

- A rapidly increasing degree of sophistication, especially with respect to control system and sensors (e.g. machine vision); and

- The integration of industrial robots with other computer-controlled manufacturing equipment. This will require extensive development work in the areas of systems control, interface and communication.

To retain its position in the technological forefront, a company must make very heavy investments in research and development (R and D). In the United States it has been estimated that in 1982 the total sum spent on R and D by the American robot manufacturers amounted to $26.5 million, or 19 per cent of the value of total shipments. 10/ Few other industries are likely to have such a high degree of R and D intensity. 11/

10/ Competitive position of U.S. producers of robotics in domestic and world markets. USTIC Publication 1475. United States International Trade Commission, Washington D.C., December 1983.

11/ Business Week, 9 July 1984. Average R and D spending by United States industry in 1983 amounted to 2.6 per cent of sales. The industry with the highest spending was the semi-conductor industry with 8.3 per cent.

Owing, on the one hand, to the fact that the robot industry is an emerging industry having strategic importance for the development of total industry and, on the other, to the large size of the necessary R and D investments, most ECE Governments have set up specific R and D programmes for robot development (see also chapter VIII). These include grants, low-interest loans, special tax deductions etc. Government R and D spending in the United States in 1982 is estimated to have amounted to over $30 million (Department of Defense $26.7 $26.7 million; National Science Foundation $4.5 million; National Bureau of Standards $1.2 million). 10/ 12/ Thus, in 1982, R and D support provided by the Government exceeded the sum spent for that purpose by industry itself.

Judging from United States data, capital investment of robot manufacturers is considerably lower than their R and D investment. In 1982, capital investment amounted to $12.4 million, that is, less than 50 per cent of R and D investment. 10/

A third type of investment concerns sales and service organization and marketing, which, if it is undertaken on a wide international basis, is extremely costly. Only the very large robot manufacturers have financial and manpower resources to engage in such investment. Smaller and medium-sized firms are therefore usually obliged to concentrate either on regional markets and/or joint-venture or licensing agreements with foreign firms (this topic is given further attention below).

Profitability

Although the robot market is expanding very rapidly 13/ most robot manufacturers are showing either very small profits or losses. 6/ 9/ 10/ 12/ 14/

(a) The large investment requirements, as discussed above; and

(b) The large number of firms (both start-ups and large enterprises) entering the robot business, and creating excess supply and very strong price competition.

An **indication** of the problem of excess supply may be found in data showing the degree of capacity utilization. In 1981, capacity utilization of American robot manufacturers was 55 per cent. 10/ In 1983 it had declined to 48 per cent. In Japan some manufacturers have postponed expansion of plant capacity owing to the presence of excess supply.

Amongst American robot manufacturers, in 1982, the median loss as a share of net sales amounted to 49 per cent. 10/

12/ J. Baranson, Robots in manufacturing - key to international competitiveness.

13/ Fuji Bank Bulletin reports that the production of industrial robots has grown by nearly 50 per cent per year since 1979.

14/ L. Conigliaro, Trends in the robot industry (revisited): where are we now? Proceeding of the 13th International Symposium on Industrial Robots, Volume I, Chicago, April 1983.

As a result of the low profits currently being experienced by robot manufacturers, the following tendencies may be observed:

- In the market-economy countries, venture capital institutions show reluctance to invest in robot firms, compared with a few years ago;

- Many of the small independent robot firms are being acquired by larger firms, the motives including the acquisition by the latter firms of technology, know-how and personnel (as an alternative to making their own R and D investment) as well as market shares; and

- Despite low profits, robot manufacturers belonging to large enterprises tend to be supported financially by the parent company because the availability of in-house robot technology is a strategic asset which leads to spill-over effects on most manufacturing activities in an enterprise.

Foreign trade

The international trade pattern varies considerably between robot manufacturing countries. At one end of the scale there are countries like Sweden and Norway, with very high export ratios. In 1982, the Swedish robot manufacturers had an export ratio of 93 per cent. 15/ One Norwegian robot manufacturer, either through direct sale or licensing, supplied some 80 per cent of all paint-spraying robots in the world.

At the other end of the scale there is Japan, which despite being the world's largest robot manufacturing country, still has a very low export ratio: in 1981 it amounted only to some 3 per cent. 12/

Major robot producing countries with moderate to large export ratios are Italy, the Federal Republic of Germany and the United States. In the latter country, the export ratio varied from 32 to 20 per cent between 1979 and 1983 10/ Table 67 provides information about the export and import of industrial robots in 1982 for the United States and Sweden.

The following pattern of trade in industrial robots between various regions can be observed:

United States: About 95 per cent of exports are placed on the west European market. During the period 1979-1983, imports, measured in terms of value, originated from the following countries: Japan (56 per cent); Sweden (13 per cent); Norway (11 per cent); Federal Republic of Germany, United Kingdom and Italy combined (20 per cent). 10/

____

15/ Production and use of industrial robots in Sweden in 1982. Computers and Electronics Commission, Ministry of Industry, DsI 1983: 1, Stockholm 1983.

<u>Table 67</u>. Production, export, import and domestic supply of
industrial robots in the United States and Sweden
in 1982

(Millions of United States dollars)

Country	Production	Export	Export ratio a/	Import	Domestic supply a/	Import ratio a/
United States	142.8	20.3	14%	15.1	137.6	11%
Sweden b/	47	43.4	92%	4.5	8.1	56%

<u>Source</u>: See foot-notes 10 and 15 above.

a/ Export ratio: Exports in percentage of production
    Domestic supply: Production + import - export
    Import ratio: Imports in percentage of domestic supply

b/ Conversion rate: $US 1 = 8 Swedish kronor.

<u>Western Europe</u>: Exports are mainly directed towards other west European countries
and the United States, with some going to east European countries and Japan.
Imports come mainly from west European countries and the United States. Imports
from Japan consist of both direct imports and licensing and joint ventures.

<u>Eastern Europe</u>: Export and import activities are mainly contained within the
region. Imports also come from Japan and west European countries.

<u>Japan</u>: Exports go to the United States and to western and eastern Europe, listed
in descending order. Imports, the quantity of which is marginal, originate
mainly from the United States, Norway (under licence) and Sweden.

    Upon examining the mode of trade in industrial robots, the following
patterns may be distinguished:

    (a) Only the more important robot manufacturers, affiliated to large
international enterprises, have established international sales subsidiaries;

    (b) Amongst this group of robot manufacturers, only a few have established
foreign robot manufacturing subsidiaries. In this context the major companies
are the following:

<u>Robot manufacturer</u>:	<u>Manufacturing subsidiaries in</u>:
ASEA	United States, France, Spain and Japan
Unimation (Westinghouse)	United Kingdom
Fujitsu Fanuc	United States (joint venture with General Motors)
Cincinnati Milacron	United Kingdom

    (c) Besides undertaking some direct export sales, most of the smaller and
medium-sized robot manufacturers export either through joint ventures or
sales/manufacturing licensing agreements with foreign firms. The cost of

establishing sales and service subsidiaries on international markets are seldom justified for this group of manufacturers, taking into account both the strong competition and the low sales volume. When discussing trade in the robot industry, it is necessary to consider not only the volume of direct sales but also the transfer of royalty payments derived from licensing agreements. As will be seen below, there is a very dense network of joint ventures and licensing agreements amongst the world's robot manufacturers.

<u>Specialization versus generalization</u>

Taken broadly, robot manufacturers can be broken down into the following three groups:

- <u>Manufacturers of special purpose robots</u>. Robots are developed for a single or only a very limited number of applications, e.g. for assembly, for servicing machine tools etc. Within this group, there is variety as regards the degree of versatility, ranging from very dedicated robots developed for a single machine tool of a particular brand to robots which can be applied to different machine tools carrying out one or a limited number of operations;

- <u>Manufacturers of general purpose robots</u>. The robots are designed in such a way that, after retooling and re-programming, they can be used in a variety of applications, e.g. spot welding, arc welding, assembly, servicing of machine tools etc.; and

- <u>Manufacturers of value-added robot systems (turnkey vendors)</u>. Manufacturers included in this group might or might not produce basic robot units. Their primary activity is to supply and install tailor-made robot systems at the site of end use. In these systems, the robot is a component - albeit the basic one - to which are attached the following ancillary pieces of equipment, manufactured by the turnkey vendor: grippers, tools and process equipment (e.g. welding equipment); fixtures and material-feeding devices; sensors (e.g. vision systems); software; interface and communication units to other equipment; host computers etc. As the total cost (to the end user) of installing a robot system can vary from 175 to 500 per cent of the initial robot cost, this type of manufacturing can reach a volume comparable to that of robot production. <u>14</u>/ <u>16</u>/

Besides taking a decision regarding the type(s) of robots he will produce, a manufacturer must also decide on

- The type(s) of application(s) on which he should concentrate; and

- The degree of sophistication of the robots and robot systems (low-cost versus high-cost systems).

The decision of a manufacturer when making his selection amongst the above-mentioned strategies is generally dependent upon his financial, technological and manpower resources as well as on the existence of international subsidiaries of the parent company for marketing purposes. Within this framework each strategy

____

<u>16</u>/ <u>Competitive position of U.S. producers of robotics in domestic and world markets</u>. USTIC Publication 1475, United States International Trade Commission, Washington D.C., December 1983.

is evaluated with respect to:

- Investment costs: R and D, capital investment and marketing;

- Market size and market growth; and

- Competition.

As shown earlier, investment cost is a very heavy item in the cost picture of a robot manufacturer, especially if all three elements listed above are included. One would therefore be tempted to conclude that the obvious strategy to select is the manufacture of general-purpose robots, with a view to achieving economies of scale with respect to both R and D and manufacturing costs. However, the marketing of general-purpose robots intended for a multitude of applications requires extensive investment in process knowledge and the development of peripherals and accessories for each type of application. Thus, the economies of scale that might be achievable with respect to the basic robot unit might therefore be lost owing to the increased costs of the development of each process application. This can of course be offset if the manufacturer has the capability of effecting large-scale production for each application area, a situation likely to occur only if the manufacturer has a strong international market.

In view of the above, the production of general-purpose robots will probably be mostly confined to large manufacturers with international outlets whilst smaller and medium-sized independent manufacturers will mainly concentrate on the manufacture of either special-purpose or value-added robot systems.

Licensing and joint venture versus in-house development and
direct international sale

As discussed above, there are two prerequisites - which are to a considerable degree contradictory - for achieving success in the robot business. On the one hand, there is the need for heavy R and D investment, which in turn provides incentives for internationalization in order to achieve a larger production volume over which the R and D cost can be distributed. On the other hand, as internationalization requires large investment in marketing and the establishment of service and installation organizations, it might justify concentration on certain regions or certain market segments.

There is, however, a third approach, which most robot manufacturers have adopted: the achievement of economies of scale through partnerships, joint ventures or licensing agreements with various companies on different markets. For the licensor the advantage lies in the fact that his R and D costs will be partly reimbursed without his needing to invest in foreign market organizations (provided of course that the licensee is successful). The licensee has the advantage of not having to invest in R and D for robot development and in being able to concentrate on value-added activities and marketing.

The importance of international co-operation in the robot area is clearly illustrated in figure 13a-13f. From this figure some further general conclusions can be drawn:

(a) Most licence agreements are drawn up between west European or Japanese manufacturers (licensors) and American manufacturers (licensees). At the same time, several west and at least one east European manufacturers have purchased licensing rights from Japanese manufacturers. There are some examples of licensing agreements signed in the direction opposite to that described above.

These include the following:

Licensor	Licensee
Unimation (United States)	Kawasaki Heavy Industries (Japan) Nokia (Finland)
Trallfa (Norway)	Kobe Steel (Japan)
DEA (Italy)	Amada Engineering (Japan)

In this context note should also be taken of the very close co-operation in the development of robots which takes place amongst member countries of the Council for Mutual Economic Assistance (CMEA).

(b) Some Japanese, American and west European manufacturers have signed licensing agreements with manufacturers in eastern Europe, for instance in Bulgaria.

(c) In many cases the co-operation is extended to take the form of a joint venture, with or without shareholding. The major alliances in this respect are:

General Motors (United States)/Fujitsu Fanuc (Japan)
Siemens (Federal Republic of Germany)/Fujitsu Fanuc (Japan)
Cincinnati Milacron (United States)/Dainichi Kiko (Japan)
Machine Intelligence Corp. (United States)/Yaskawa Electric (Japan)
Ransburg Corp. (United States)/Renault Automation (France)
Bendix (United States)/Comau (Italy)
IBM (United States)/Ilsag (Italy).

(d) Upon examining co-operation amongst robot manufacturers in western Europe, G. Junne and R. van Tulder found it to be rather insignificant. 17/ Most of the agreements signed are either between very small firms or between a corporation and its European subsidiaries.

Figure 13a-13f gives examples of acquisitions and joint ventures within individual regions and countries.

<u>Illustration of structure and extent of international co-operation in the robot area</u>

Figure 13a-13f summarizes some of the major international agreements on co-operation in the robot area. The figure purports to depict the degree as well as the forms (sales/manufacturing licensing, joint ventures, acquisitions) of co-operation that have been established. It also gives an indication of the countries and regions between which co-operation schemes have been set up.

17/ G. Junne and R. van Tulder. <u>European multinationals in the robot industry</u>. University of Amsterdam, January 1984. (Draft study sponsored by the Institute for Research and Information on Multinationals - IRM).

When assessing the information provided in figure 13a-13f, due attention should be paid to the following circumstances:

- The information is based on a large number of sources: books, reports, periodicals and newspaper articles. As the information in the sources referred to has not always been verified, there are probably some items of information that are incorrect. For this reason the information in figure 13a-13f should serve only as an indication of international co-operation.

- The picture is not complete. There are probably some agreements which are not accounted for and others that have been terminated. As new co-operation constellations are constantly being formed, the figure can give no more than a general picture of the situation 1983/84.

- Foreign industrial robot-sales subsidiaries are only marginally accounted for.

VIII. SOME GENERAL TECHNICAL, ECONOMIC AND SOCIAL IMPLICATIONS
RELATING TO THE INTRODUCTION OF ROBOTS

The present decade is characterized by the rapid diffusion of microelectronics in industry which will lead to increases in labour productivity and to significant changes in the structure of jobs. That is to say, in the field of robotics, where today the share of electronic-control-system costs is at least 50 per cent, various jobs may be expected to disappear, while some new jobs will be created, for instance in the manufacture and design of automatic control devices and in servicing those devices. Some general conclusions on the impact of microelectronics on the evolution of the employment structure 1/ apply very well to the present diffusion of robots. Physical labour is diminishing and jobs will slowly acquire a more supervisory and guarding nature. Jobs will become more abstract, creating a need for new education and training approaches, even at worker and operator levels.

The linkage of mechanics with electronics is sometimes called "mechatronics". 2/ It deals with the application of advanced technologies, such as microelectronics, to achieve greater efficiency in the design, production and operation of machinery. Robotics is believed to be the leading mechatronic principle. Some authors 3/ are using the term "digital revolution", which is based on the merging of microcircuits with pulsed-signal technology. In other words, robotics seems to be a typical example of the natural convergence of automatic control and information technology in the field of factory automation.

Current and prospective issues in technology policies are continuously studied by the Senior Advisers to ECE Governments on Science and Technology. In a note issued by the secretariat on the impact of microelectronic-based technologies in industry 4/ the important role of robotization was underlined. Opinions on future developments and their socio-economic implications as well as some generalized policy considerations included in the above-mentioned note are certainly valid for the application of industrial robots. Close co-operation between the Senior Advisers and the Working Party on Engineering Industries and Automation will continue, e.g. in the preparation of the ECE Seminar on Industrial Robotics '86 (see chapter I).

The following areas of benefits (economic effects, savings, etc.) justify the growing use of industrial robots:

1/ J. J. A. Volleberg, "Micro-electronics and the Evolution of Society". Microprocessing and Microprogramming, No. 7, 1981.

2/ "Japan Corporate Strategies for the 80's". Business Week International, 19 July 1982.

3/ W. A. McAdams, "Into the Digital Revolution". International Electrotechnical Commission (IEC) Bulletin, vol. 16, No. 75, May 1982.

4/ Perspectives on the development and diffusion of selected microelectronic-based technologies in industry (SC.TECH./R.114).

- Increase in throughput and productivity;

- Fulfilment of high-quality demands;

- Labour-cost savings, including replacement of workers in monotonous and hazardous jobs;

- Other cost savings (e.g. energy and materials).

The above-mentioned groups of benefits are closely related, and include the following aspects: 5/

High-speed operation (faster throughout times), reduced value of work-in-progress, quick automatic resetting of equipment and tool changing, availability of equipment 24 hours (three shifts) per day and, high reliability (minimum breakdowns) seem to be the main factor contributing to the increase of total gross output and productivity growth at factory level. For example, the mass introduction of robots in the USSR (watch assembly, electrical machinery, hot and cold pressing etc.) increased labour productivity by an average factor of six. 6/ Improved utilization of capital assets may also supplement economic benefits considerably:

- As the manual jobs and operations in which robots replace workers are naturally those that are most hazardous, physically and mentally demanding, tiring etc., it is obvious that the introduction of robots will result in greater safety and accuracy and a decrease in rejects: the robot is generally more consistent on the job and its use will significantly increase net output.

- The prime issue in justifying the introduction of robots is labour displacement, either by transferring workers and operators from hazardous working conditions to more convenient jobs or simply by saving labour costs: on an average, one robot can replace two to three workers.

- The introduction of a robot significantly helps to meet the demand for higher product quality and a reduction in the time lapse between receipt and confirmation of the customer's order and its filling. "Flexible" order books may be rescheduled quite frequently in order to reduce the value of work-in-progress and to introduce automated inventory control of semi-finished and finished products. It is also possible significantly to reduce eventual customer claims. In other words, manufacturing processes equipped with robots seem more competitive in the market, as has already been proved in the automotive industry: car models based on computerized design, engine testing, robotized body welding and spray-painting create customer confidence and meet his wish to own a reliable product with customized options.

5/ J. F. Engelberger, Robotics in Practice. London, Kogan Page, 1980.

6/ ECE/ENG.AUT/9, para. 20.

-      Among the other economic effects, those concerning raw material and
       energy savings are most important. Automated material-requirement
       planning provides for optimum stock levels of various raw materials,
       assembly parts, semi-finished products etc. Consumption of various
       forms of energy, especially stable consumption of electric energy
       within the automated manufacturing process, can also be controlled
       and effectively contributes to the financial justification of robots.
       In addition, various organizational structures and processes may
       be simplified, such as maintenance, dispatching, transport,
       accounting, invoicing and other financial operations.

On the other hand, robot costs consist of:

-      The purchase price of the robot installation, including accessories,
       peripherals etc;

-      Maintenance and periodic overhaul;

-      Operating costs;

-      Depreciation (return rate of investment); and

-      Personnel costs including training.

The purchase price of a robot varies from $US 5,000 to 100,000 depending
on the number of articulations, the working space, the loading capacity, the
number and sophistication of control functions and so on.

-      Special tooling might include various conveyors, transformers,
       welding guns and special controlling equipment needed for the
       interface with the workplace. Additional installation costs are
       due to plant layout changes caused by the introduction of the robot
       system.

-      Any well-implemented robot system working approximately two shifts
       needs regular maintenance, a periodic complete overhaul, and
       immediate correction of any malfunctioning or breakdowns. The total
       annual cost of this maintenance comes to 10 per cent of the
       acquisition cost, but depends also on the job and environment demands:
       maintenance costs in a foundry are higher than for simple loading
       operations.

-      The operating costs include power requirements: energy costs are
       usually higher for automated manufacturing than for manual control
       and other auxiliary operations; they are not, however, a major
       expense in comparison with the purchase price.

-      Robots have a useful life and it is ordinary practice to depreciate
       the investment over this useful life. Since a robot tends to be a
       piece of general-purpose equipment, it is evident that a 10-year
       life span is conservative. When a company purchases a robot
       priced at $US 60,000, replacing per shift one worker with approximate
       annual wages of $US 20,000, payback is three years for a one-shift
       operation, 1.5 years for a two-shift operation, and so on.

The successful introduction of robots into an existing plant (as opposed to a newly built factory) depends on the "readiness" of those involved to accept new technology and adapt to new methods of work. Costly training of personnel, including management, and the modification of qualification requirements of employees (electronic engineers, programmers) is often necessary.

Figure 14 shows the relation of unit manufacturing cost to production volume for different types of operations.

The distribution of labour within a typical manufacturing process in engineering industries may be seen in table 68. For example, in the automotive industry, which is the major robot user, the most labour consuming processes are preferably being automated, i.e. basic manufacturing (spot welding, machine loading) and finishing operations (paint spraying, 7/ assembly).

Table 68. Labour distribution by manufacturing processes

Manufacturing process	Percentage	
Casting	4.8	
Forging	1.3	
Total semi-finished products		6.1
Machine tools	32.8	
Forming	6.9	
Welding	6.6	
Heat treatment	1.3	
Total basic manufacturing		47.6
Surface treatment	6.4	
Assembly and manual operations	36.3	
Total finishing		42.7
Other	3.6	3.6
Total percentage	100.0	100.0

Source: V. Krenek, Handbook on Automation in Engineering, chap. 7.5 Prague, SNTL Publishing House, 1970.   (In Czech)

Another source (see table 69) gives figures on labour distribution in four major durable-goods industries in the United States. It also points to the intensive labour use in assembly processes and in the fabrication of precision parts, namely in metalworking machinery.

7/ A painting robot uses from 20 to 30 per cent less paint than the worker it replaces. Business Week International, 19 July 1982.

Table 69. Labour distribution by selected manufacturing
processes in the United States

(Percentage)

Type of job	Labour distribution			
	Motor vehicles	Farm machinery	Radio and TV receivers	Metalworking machinery
Non-production workers	39.4	42.7	52.7	32.2
Production workers:				
- Assembly	33.4	30.2	23.9	11.0
- Parts fabrication	16.6	20.2	8.4	50.2
- Inspection	9.0	5.5	13.3	5.4
- Other	1.6	1.4	1.7	1.2
Total	100.0	100.0	100.0	100.0

Source: United States Census, 1970.

A study published by the International Labour Organisation (ILO) 8/
describes the social implications of introducing information technology.
Long-term trends seem to suggest that we may be facing a transition towards a
society with no need for the full-time use of its entire potential labour
force. Some recent major high-technology ventures have recognized the vital
role of training: 9/ sophisticated manufacturing systems are only worth as much
as the people who manipulate them.

A poll was organized in Japan to obtain opinions on the introduction of
robots and office automation and responses were received from 68 companies
(management) and 38 unions (labour). The questionnaire contained nine groups
of questions and the most frequently recurring responses (exceeding 25 per cent)
are shown in tables 70, 71 and 72.

Another survey among robot users was undertaken in the United Kingdom by
Imperial College; 10/ from 403 questionnaires sent out, 115 were returned
completed. The survey results pointed out the following main reasons for
deciding to install robots, in order of priority:

(a)   To save labour;

(b)   To replace people working in dangerous or obnoxious conditions;

(c)   To provide a more flexible production system;

(d)   To achieve more consistent quality control;

8/   J. Rada, The impact of microelectronics. ILO, Geneva 1980.

9/   K. H. Ebel, "The microelectronics training gap in the metal trades".
International Labour Review, vol. 120, No.6, 1981.

10/   "Robotics in the United Kingdom", The Industrial Robot, March 1981.

(e)   To increase output; and

(f)   To overcome a shortage of skilled labour.

Table 70. Opinions on the introduction of robots and office automation
(Percentage of responses)

Group of questions	Response	
Labour expectations	Reduced monotonous labour	68.4
	Reduced exhausting labour	50.0
	More scope for developing abilities	31.6
Labour reservations	More people who cannot adapt	65.8
	Increased alienation from work	52.6
Anticipated impact on employment	Surplus workers hired by new businesses (management)	39.7
	Surplus workers transferred (management)	35.3
	Unemployment of young workers (labour)	42.1
Management's expectations	More effective utilization of human resources	75.0
	More effective responses to new technology	41.2
	Reduced personnel costs	32.4
How management will deal with excess personnel	Transfer to other sections	45.5
Degree of employee adaptability	Fully adaptable (management)	26.5
	Problems in some areas (management)	42.6
	Problems in some areas (labour)	34.2
	Problems for older workers (labour)	50.1
Management's investment policy	Gradual investment	72.1

Source: Nikkei Business Editorial Department, "Attitudes toward automation",
Economic Eye, June 1982.

Table 71. Development of activities: 10-year outlook
(Percentage of responses)

Activity	Assessment of progress	Management	Labour
Production	Accelerating	26.5	26.3
	Gradual	41.2	47.4
Research	Accelerating	23.5	36.8
	Gradual	41.2	21.1
Management	Rapid	19.1	28.9
	Accelerating	39.7	44.8
	Gradual	36.8	23.7
Sales	Accelerating	36.8	39.4
	Gradual	42.6	34.2

Source: As for table 70.

Table 72. Changes in number of employees: 10-year outlook of management
(Percentage of responses)

Employment within activity	Response	Investment policy	
		Aggressive	Gradual
Production	No change Decrease No response	43.8 18.8 25.0	14.3 38.8 22.4
Research	Increase	75.0	55.1
Management	No change Decrease	49.9 31.3	24.5 57.2
Sales	Increase	74.9	63.3

Source: As for table 70.

In this connection, it is interesting to note a view resulting from a study undertaken by the Joint Economic Sub-Committee on the Monetary and Fiscal Policy in the United States, 11/ i.e. that the introduction of robots will increase the numbers of available jobs and improve living standards, rejecting the common belief that robotics is eliminating jobs.

In order to illustrate a typical investment calculation, two examples of installations in Sweden are described below. 12/

Example I

Type of investment: Investment in an industrial robot service to a machine group consisting of two CNC-lathes, a measuring station and a conveyor.

Investment objects: The two lathes were already in use but served manually. The new investment objects were the robot, accessories, a measuring station and a conveyor.

Operation cycle: The robot picks up a part from a buffer storage, puts it in lathe I. At the same time it takes out a part that has just been machined (the robot has a double gripper). This part is then moved to lathe II, where the robot also takes out a machined part. This part is then moved to a measuring station and if it is accepted it is put on the conveyor.

This is an example of a well-balanced and fully automatic machining cell. The cell was planned for a two-shift operation but can also work unmanned during the night. Thus, the cell has a high degree of utilization.

11/ Industry Week, 3 May 1982.

12/ J. Carlsson, "Production and Use of Industrial Robots in Sweden in 1982". Computers and Electronics Commission, Ministry of Industry, Stockholm.

Investment calculation: The detailed investment calculation is shown in table 73.

Table 73. Investment calculation for example I
(Thousands of Swedish Kronor)

Indicator	Ex ante (Estimate 1976) a/		Ex post (Actual 1979) a/	
Investment cost (of which the robot = 500)		1 500		1 600
Savings through:				
- Reduced labour		(375		(340
- Lower interest on work in progress		(100		(100
- 15-per cent increase in productivity	560	(100	625	(200
Maintenance costs		((-15		((-15

Source: Computers and Electronics Commission, Ministry of Industry, Stockholm.

a/ Production volume: 150,000 parts/year.

Pay-off time:

Ex ante: $\dfrac{1\ 500}{560}$ = 2.68 years.

Ex post: $\dfrac{1\ 600}{625}$ = 2.56 years.

Internal rate of return: It is assumed that the industrial robot can work for 10 years and after that "the rest value" is 0.

$$\sum_{t=1}^{n} \frac{a_t}{(I+p)^t} = I$$

where I = Investment cost

a_t = Savings year t

p = Internal rate of return (per cent)

n = Number of years.

Ex ante: $\displaystyle\sum_{t=1}^{10} \frac{560}{(I+p)^t}$ = 1 500 → p = 36 per cent.

Ex post: $\displaystyle\sum_{t=1}^{10} \frac{625}{(I+p)^t}$ = 1 600 → p = 38 per cent.

<u>Example II</u>

<u>Type of investment</u>: Investment in two industrial robots to service a
machine group consisting of two broaching machines and a drilling machine,
connected with a conveyor. The robots were installed at Saab-Scania in a
section where connecting-rods for diesel engines are produced. Except for
a paint-spraying robot, this was the first robot installation at this plant.

The motives for the investment were:

-    To increase productivity; and

-    To improve the working conditions and thereby reduce the degree
     of labour turnover.

This production section was characterized by difficult working conditions:

-    Heavy lifting: each rod weighs 7 kg and was handled six times
     and 60 rods were produced per hour, i.e. a worker lifted
     2,500 kg/hour; and

-    The broaching machines required a lot of cutting oil, as the rods
     are very greasy and unpleasant to handle manually. The oil
     stains caused diseases among the workers.

<u>Investment objects</u>: Two robots and a conveyor.

<u>Operation cycle</u>: Robot I picks up the rods from the conveyor and puts
them into machine I. There each rod is placed in four fixture positions
and into two intermediate buffer storages. When the machining is completed
the rods are placed into a third intermediate buffer storage. There they
are picked up by robot II which puts the rods into machine II, in two
different fixture positions. When the machining is completed, robot II puts
the rods on to the conveyor and gives a signal to the conveyor to move
forward.

<u>Investment calculation</u>: The detailed investment calculation is shown in
table 74.

Table 74. Investment calculation for example II
(Thousands of Swedish kronor)

	Ex ante (Estimate 1976)	Ex post (Actual 1978)	Ex post (Actual 1979)
Investment cost			
- Two robots)			
- Conveyor)			
- Other transport equipment)			
and accessories)	1 079	1 079	1 079
Installation costs	190	340	340
Subtotal	1 269	1 419	1 419
Saving through delayed investment for capacity increase (10% of 2 000)	-200	-200	-200
Total investment costs	1 069	1 219	1 219
Savings:			
- Labour cost	289	396	460
- Oil consumption	34	34	40
Total savings	323	430	500

Source: As for table 73.

Pay-off time:

$$\text{Ex ante:} \quad \frac{1\ 069}{323} = 3.3 \text{ years}$$

$$\text{Ex post (1978):} \quad \frac{1\ 219}{430} = 2.8 \text{ years}$$

$$\text{Ex post (1979):} \quad \frac{1\ 219}{500} = 2.4 \text{ years}$$

Number of workers before and after the installation of the robot

- Before robot installation (Ten men were assigned to this station: working conditions were difficult and absenteeism heavy)   7.6

- Estimated number of workers after robot installation (Calculus ex ante)   4.1

- Actual number of workers required after robot installation
  - 1978   2.7
  - 1979   2.5

Time-table for the robot installation

	Date	Capacity (Percentage)
- Project start: feasibility study	January 1977	
- Investment decision	January 1978	
- Installation	June-August 1978	
- Start of production	14 August 1978	60

	Date	Capacity (Percentage)
- Programme change 1	31 August 1978	80
- Programme change 2	7 September 1978	92
- Programme change 3	1 October 1978	97
- New production target 13/	31 December 1978	110

Another study on the profitability of industrial robots was undertaken in the Federal Republic of Germany. 14/ The pay-off times were calculated for different applications of robots, as shown in table 75. The calculation was based on assumptions of the following annual growth rates: wage costs 6.5 per cent, other running costs 4.0 per cent and interest rate 9.0 per cent.

Table 75. Profitability of industrial robots by area of application in the Federal Republic of Germany

Area of application	Percentage of total purchase costs		Pay-off time b/ (years)
	Industrial robot	Peripherals, accessories a/	
Loading and unloading of machine tools	39-53	22-29	1.75-7.25
Loading and unloading of steel-shaping machines	47-54	23-28	1.50-2.25
Loading and unloading of plastic-injection-moulding machines	60-61	13-18	1.50-4.25
Spot welding	38-42	38-41	2.00-5.00
Track welding	33	46	2.25-5.50
Paint spraying	56-71	17-32	4.25-10.00
Finishing of mouldings	65	21	2.50-6.50
Assembly processes	41-53	30-40	3.00-10.00

Source: Microelectronics, Robotics and Jobs, ICCP Series No. 7, OECD, Paris 1982.

a/ Hardware costs associated directly with the functioning of the robot (excluding installation costs, planning and programming, training of personnel and other costs).

b/ Depending on the number of shifts per day.

13/ The machine group was planned for a two-shift operation. Owing to increased demand, the system was run unmanned during parts of a third shift.

14/ V. Volkholz, "Trends in the use of industrial robots in the 1980s - the case of the Federal Republic of Germany". Microelectronics, Robotics and Jobs, ICCP (Information, Computer, Communication, Policy) Series No. 7, OECD, Paris 1982.

A possible way of assessing the efficiency of the use of industrial robots at the micro-economic level was described by Katzarov and Christov. 15/ It is based on the approach introduced by Benedetti (1977) 16/ and Anguelov (1981), 17/ i.e. on the following inequality:

$$T_M \cdot C_{LS} > T_R \cdot C_{RS} + T_O \cdot C'_{LS} + C_{REP} + C_{PLAN} \qquad \ldots (1)$$

where

T_M =   Human production time per workpiece

C_{LS} =   Cost of worker's time per second

T_R =   Work-cycle time for robot per workpiece

C_{RS} =   Fraction of the robot's cost allocated to each workpiece per second

T_O =   Operator intervention time (seconds) for each workpiece produced by robot

C'_{LS} =   Cost of one second of operator time

C_{REP} =   Cost of reprogramming the robot for a new series per workpiece

C_{PLAN} =   Cost of planned maintenance per workpiece.

According to (1), the introduction of the robot is justified if the inequality is fulfilled. Variables T_M, C_{LS} and C_{RS} are evaluated within the computer programme ROBO-DISCO. 15/

A different approach to the assessment of the economic efficiency resulting from the introduction of industrial robots was made by Granicki, 18/ at both the micro-economic (2) and macro-economic levels (3):

$$E_{MI} = \frac{P - C_p}{\sum_i I_i \, f \, (r + D_i)} \qquad \ldots (2)$$

$$E_{MA} = \sum_j \left[\frac{P - C_p + n \, (T + S)}{\sum_i I_i \, f \, (r + D_i)} \right] \qquad \ldots (3)$$

where   P   =   Annual output in monetary units after the introduction of IR

       C_p =   Annual production costs after the introduction of IR (including various taxes and payments)

15/ B. B. Katzarov and T. M. Christov. "Preliminary Assessment of the Profitability of Industrial Robot Application". IIASA, Laxenburg, August 1982.

16/ M. Benedetti, "The Economics of Robots in Industrial Applications". The industrial robot, vol. 4, No. 3.

17/ A. S. Anguelov et al., "Robotization and Social Labour Productivity Growth". Technica, Sofia, 1981 (in Bulgarian).

18/ J. Granicki, "Evaluation of economic efficiency of use of industrial robots in Poland". Paper presented to ECE Seminar on Development and Use of Industrial Handling Equipment, Sofia 1979 (ENGIN/SEM.5/R.27).

I_i = Investment (purchase) cost of i-th equipment

f = "Freezing" coefficient

r = Interest rates

D_i = Depreciation rate of i-th equipment

n = Average number of shifts

T = Additional profit achieved by transfer of workers to other jobs

S = Social savings due to better working conditions

i = Index of IR installation within enterprise

j = Enterprise index.

It is obvious that both the values E_{MI} and E_{MA} will increase with increasing output and decreasing production and investment costs. In any case, the success of this method depends on the availability of the relevant data.

For quick management orientation (short-term frame) the payback period calculation formula usually gives a feasible estimate: 19/

$$P = \frac{I}{L-E} \qquad \ldots (4)$$

where

P = Payback period (in years)

I = Total capital investment in robot and its accessories (value)

L = Annual labour costs replaced by the robot (value per year)

E = Annual maintenance costs (value per year)

As the life span of a robot is of many years, it seems advantageous to evaluate its long-term effect in terms of return on investment, as proposed in the same source: 19/

$$R = \frac{S.100}{I} \qquad \ldots (5)$$

where

R = Return on investment (in per cent)

S = Annual savings from the robot installation (value per year)

The annual savings S might be expressed:

$$S = L-C \qquad \ldots (6)$$

19/ J. F. Engelberger, Robotics in Practice. American Management Association, New York, 1980.

where

$\quad$ C $\quad$ = $\quad$ Annual robot costs (value per year)

and

$\quad$ C $\quad$ = $\quad$ $\dfrac{I}{N-E}$ $\qquad\qquad\qquad$ ... (7)

where

$\quad$ N $\quad$ = $\quad$ useful life of the robot (in years)

According to one source, [20] since 1961, labour costs have increased by some 250 per cent, whereas the cost of a robot has increased by only 40 per cent. The utilization of robots improves productivity from 10 to 90 per cent (depending on the type of application) and could reduce the average direct labour costs by some 60 per cent per robot hour including all operating and depreciation costs).

The economic justification for using robots and automated manufacturing was also a recurring theme of the AUTOFACT Europe Conference in 1983. The justification is based mainly on a comparison between the capital costs of the installation and the relevant cash flow benefits, including those resulting from reduced labour content. In this respect, the following three appraisal techniques were suggested:

(a)$\quad$ The payback method;

(b)$\quad$ The return on investments analysis;$\quad$ and

(c)$\quad$ The discounted cash flow (DCF) technique.

The DCF technique seems to be advantageous - given present high-interest-rates - since it takes into account the time value of money; [21] it can be evaluated in two forms, namely:

(a)$\quad$ The yield-of-rate-of-return method, where the task is to find the minimum value of rate of return (r) in the equation

$$\sum_{i=1}^{T} \frac{R_i}{(1+r)^i} - E = 0 \qquad\qquad ... (8)$$

where

$\quad$ E $\quad$ ... capital expenditure on the robot installation

$\quad$ R_i $\quad$ ... the resultant net cash flow at the end of the i-th period (i = 1, 2, ..., T)

$\quad$ T $\quad$ ... economic life of the robot (years, months)

---

[20]$\quad$ R. L. Maiette "The Use of Robotics as an Integral Part in the Planning of Your Flexible Machine Tool Cell" AUTOFACT Europe Conference Proceedings, Geneva, 1983.

[21]$\quad$ M. P. Kelly, R. J. Grieve, P. H. Lowe: "Robots: The Economic Justification". AUTOFACT Europe Conference Proceedings, Geneva, 1983.

(b)   The net-present-value (NPV) method, where the task is to fulfil
the following inequality

$$NPV = \sum_{i=1}^{T} \frac{R_i}{(1+S)^i} - E > 0 \qquad \ldots (9)$$

where S is the target rate of return for NPV analysis.

Some examples of the economic advantages gained by the use of industrial
robots in the automotive industry are mentioned in a study undertaken by
OECD: 22/

- Chrysler replaced 200 welders over two shifts by the introduction
  of 50 robots in one plant; the production rate increased from 50
  to 65 cars per hour.

- Fiat increased its number of robots by 30 per cent in order to
  increase automatic spot-welding operations by up to 98.5 per cent;
  direct labour time per car body decreased by 1.5 hours.

- The introduction of 125 robots at the Renault plant at Douai increased
  the output to 80 cars per man-year in 1982.

- The use of a robot for adhesive bonding of trunk lids at British
  Leyland increased the throughput by 100 per cent.

An interesting study on the social implications of robotics was undertaken
by The Social Research Institute (SOFI) at Göttingen and the University of
Bremen in the Federal Republic of Germany. 23/ Investigation of 13 different
workplace cases with 27 industrial robots concerned changes in various types
of strains which occurred after the installation of the robots. While both
the physical strain (mainly moving of heavy loads) and the environmental
strain (gas, oil, liquids, dust, wearing of protective clothing, vibrations,
etc) were significantly reduced, some negative results were obtained involving
increased psychic strain. Operators servicing the robots work at a fixed
rhythm and in social isolation; this includes responsible supervisory functions,
which require systematic training efforts.

With the increased use of robots in industry, the safety of workers and
operators working near the installations becomes a serious problem. History
already records that K. Urada, a maintenance worker in a Japanese factory, was
the first person to be killed by an industrial robot, in 1981. Therefore,
international co-operation in developing measures for the protection of personnel
seems to be inevitable. In this context, the efforts of General Motors in
introducing the safety sensor system Roboguard should be mentioned. 24/

22/   "The Impact of Industrial Robots on the Manufacturing Industries of
Member Countries". Draft of OECD study, Paris, October 1982.

23/   W. Wobbe-Ohlenburg, "The Influence of Robots on Qualification and
Strain". Proceedings of the International Conference on Robots in the Automotive
Industry, Birmingham, United Kingdom, 20-22 April 1982.

24/   "Protecting workers from robots": American Machinist, March 1984.

It is not the aim of the present study to review the advantages and disadvantages of various approaches adopted at the national or company level towards the training of personnel in the field of robotics. However, it seems that one of the basic conditions for success in this area might be the will and flexibility of all parties concerned, namely workers, operators and management. During the meeting of representatives of scientific and technical societies of socialist countries, held at Warsaw 25/ and representing about 12 million member engineers, it was agreed that a special support programme should be provided in the field of microelectronics and robotics. Steps towards the creation of a robot demonstration and education centre 26/ which will serve new and potential robot users have also been taken in the United Kingdom.

Special government measures to promote and encourage robot installations have been taken in many countries. 27/ In the United Kingdom the aim is to reduce the risks for companies with limited experience that want to install new types of robots in applications. 28/ As described by J. Le Quément, 29/ "most Governments support major research and development programmes, thus taking over from industrial groups, in advanced areas of research (in the United States, the IPAD programme conducted by NASA and the ICAM programme under the auspices of the United States Air Force; in Japan, the national flexible workshop programme using the laser (1977-1983) and probably beginning in 1982, a national advanced robotics programme; in the Nordic countries, the Mekoflex programme, etc)".

In France, 30/ "the foundation in October 1980 of the Advanced Automation and Robotics Group was the result of a nation-wide research policy in this area, formed by the following 10 participants: the National Scientific Research Centre (CNRS), the National Automated Production Development Agency (ADEPA), the Engineering Industry Technical Study Centre (CETIM), the National Space Research Centre (CNES), the Atomic Energy Commission (CFA), the French Petroleum Institute (IFP), the National Computer and Automated System Research Institute (INRIA), the National Bureau for Aerospace Studies and Research (ONERA-CERT), the Renault Company (RNUR), and the Télémécanique Company. Spread over 1980-1984, the ARA project is split into four main items:

(a)   General robotics,

(b)   Engineering and technology,

(c)   Advanced remote operations,

(d)   Flexible production systems".

25/ A. Grishkov, "Automation dictates the conditions". Socialisticheskaja Industrija, 24 April 1982.

26/ Trybuna Ludu, 1 July 1982.

27/ The Engineer, 6 May 1982.

28/ For some examples see SC.TECH/R.114, paras. 38-50.

29/ J. Hollingum, "Cash as you spend, robot plan". The Engineer, 11 June 1981.

30/ J. Le Quément, "The Social and Economic Stakes in International Competition in the Robotics Industry". Proceedings of the 12th International Symposium on Industrial Robots, Paris 9-11 June 1982 (IFS London).

In 1982, a five-year development programme in electronics was announced by the French Government. 31/ More than 140 billion French francs (1982) including 55 billion francs government support will be used during this period on technological innovation, production increase, intensive training and education of personnel, etc. Some 80 billion francs from this sum will be invested in telecommunications and professional electronics and at least 15 billion francs in industrial automation (including robotics). It is also believed that this programme will create 80,000 new workplaces.

The Swedish programme 32/ for the promotion of robotics and CAD/CAM is conducted by the Swedish Board for Technical Development (STU), which is an agency reporting to the Ministry of Industry. STU funds advanced R and D in universities, research laboratories and industry. During 1972-1979, STU funding of R and D in robotics and CAD/CAM amounted to about 25 million Swedish kronor. During the 1980s, STU support for R and D in the engineering industries will increase considerably. An allocation of SKr 260 million has been predicted for the period 1980/81 - 1984/85 and a large share of it will be used for long-term projects. With respect to robotics and CAD/CAM, the following long-term projects have been initiated or are planned:

- CAD/CAM: About SKr 14 million will be allocated to this programme during 1980-1985.

- Adaptive control of machine tools: This programme was started during 1981/82.

- CAD 80: A four-year joint-venture project undertaken jointly by STU, Saab-Scania and two universities. The project cost is estimated at SKr 10 million, which is to be shared by STU and Saab-Scania. The latter will also invest about SKr 3 million in order to commercialize the CAD/CAM system for small and medium-sized firms.

- Adaptive control of industrial robots: This programme was also started during 1981/82.

Another major effort for promoting robotics and CAD/CAM is an R and D agreement between STU and the Swedish Association of Mechanical and Electrical Industries. According to this agreement, the two parties will sponsor a five-year research programme (1980/81 - 1984/85) allocating SKr 46 and 48 million, respectively. During 1982 the Government of Sweden decided to establish, in co-operation with industry, four Engineering Development Centres (EDC). Each centre will include CAD/CAM facilities and their tasks will be:

- To promote R and D co-operation between industry and universities;

- To perform industrial R and D;

- To promote the diffusion of new technologies.

For the fiscal year 1982/83 the Government has allocated SKr 24 million for the establishment of the four EDCs.

31/ Le Monde, 30 July 1982.

32/ Information received from the Government of Sweden.

Support for the promotion of robotics and CAD/CAM is also available through general industrial programmes i.e. the Industrial Development Fund and the Regional Development Funds.

Special measures for promoting production and use of programmable automation have also been undertaken in the Netherlands. These include: 33/

(a)   A two-phase R and D programme which was launched in October 1982 by the Minsitry of Education and Research Policy and the Ministry of Economic Affairs. The aim of the first phase, lasting two years and with a budget of 2 million Netherlands guilders, is to bring R and D in the field of flexible automation and robotics at the Technical Universities and at the Netherlands Organization for Applied Scientific Research (TNO) up to an international level. The second phase will be industry-oriented; and

(b)   A programme to stimulate the application of flexible automation and robotics in industry which is being prepared by the Ministry of Economic Affairs. The main aspects of that programme are:

- An awareness-promoting programme, lasting two years and with a budget of 12 million Netherlands guilders, under which enterprises will be able to apply for subsidies and low-interest loans in order to invest in applications of flexible automation and robotics new in the Netherlands - so-called demonstration projects. The aim of this scheme is to accelerate the diffusion of new manufacturing methods in industry and it is expected to start during the first half of 1983.

- Following completion of the awareness-promoting programme, a programme will be considered to provide financial support for all investments in the field of flexible automation and robotics, especially in the engineering industries.

The thirty-sixth session of CMEA, held in Budapest in June 1982, 34/ reviewed government measures concerning robotics in various east European countries and discussed the effects of the introduction of microprocessing techniques. An Agreement was signed concerning specialization and co-operation in the production of industrial robots and common research and development activities up to the year 1995. Under the robotics programme, serial production will be undertaken of automated machine-complexes, instruments and control systems, based on microelectronic means; practical steps towards the implementation of this programme are already being taken in a number of countries, including Czechoslovakia. 35/

The secretariat of CMEA has submitted information on multilateral co-operation among the CMEA member countries in the field of robot technology for the consideration of the ECE ad hoc Meeting on the present study in 1983. The complete text of the CMEA contribution may be found in annex VII.

33/ Information received from the Government of the Netherlands.

34/ Ekonomitcheskaja gazeta, No. 25, June 1982 (in Russian).

35/ J. Poslt, J. Hejsek, "From programme to reality". Hospodarske noviny, No. 30, 30 July 1982 (in Czech).

The economic and social implications of robotics and special government
and non-governmental measures in this field will certainly become more important
with the introduction of intelligent robots. The first optical, tactile and
force sensing robot systems are already being developed and tested. 36/
Examples of intelligent-robot tasks include sorting and orienting different
types of parts mixed together in the same bowl (programmable bowl feeders),
loading and unloading parts from an overhead moving conveyor, recognizing
three-dimensional objects, inserting pegs into holes (in assembly automation)
and so on. The application of third generation robots will permit manufacture
without the presence of workers and operators at the workplace. 37/ This seems
to be most important for hazardous processes such as those undertaken in
extremes of high and low temperatures, in the presence of radioactivity, or
under water, etc. Another interesting investigation is taking place at the
Institute of Cybernetics at Kiev in the USSR. 38/ Voice and speech recognition
robots are being developed which "understand" more than 1,000 words.

During the last few years significant improvements have been achieved,
notably in the field of machine vision (analysis and interpretation of images).
Industry-wide sales were estimated at the level of $US 40 million in 1983,
and of $80 million in 1984. Annual industry sales are expected to reach
$1,000 million by 1990 and some $10-20 billion at the end of century. 39/ The
most promising application area of machine vision seems to be the automated
manufacturing of discrete parts in engineering industries (table 76). A typical
structure of a machine vision system is shown in figure 15.

36/ "The robot hears the command". Izvestija, 26 November 1980 (in
Russian).

37/ "Automated Factories of the Future are not Castles in the Air" and
"Looking to sensing to restore a tarnished productivity image". Sensor Review,
January 1981.

38/ E. Popov, "Levels of autonomy". Socialisticheskaja industrija,
18 November 1980 (in Russian).

39/ G. Schaffer, "Machine vision: a sense for computer integrated
manufacturing". Special report 767, American Machinist, June 1984.

Table 76. Review of current machine vision applications

Applications	Feature-measurement capabilities					Image Model		Performance		
	Image shapes	Distance (range)	Orientation	Motion	Surface shading	Two dimensions	Three dimensions	High resolution	High speed	High discrimination
Inspection:										
- Dimensional accuracy	X					X			X	
- Hole location and number	X					X			X	
- Component verification	X		X			X			X	
- Component defects	X					X			X	
- Surface flaws					X	X				X
- Surface-contour accuracy		X					X	X		X
Part identification:										
- Part sorting	X					X			X	
- Palletizing	X		X			X			X	
- Character recognition	X					X			X	
- Inventory monitoring	X					X			X	
- Conveyor picking, no overlap	X		X			X			X	
- Conveyor picking, overlap	X		X		X	X		X		X
- Bin picking	X	X	X		X		X	X		X
Guidance and control:										
- Seam-weld tracking		X	X		X		X	X		X
- Part positioning	X	X	X				X	X		
- Processing/machining	X	X	X				X	X		
- Fastening/assembly	X	X	X				X	X		
- Collision avoidance	X	X	X	X			X		X	

Source: See footnote 39/.

Another emerging manufacturing concept, closely related to robot development, is the field of laser machining, whose applications are now beginning to increase rapidly. 40/ According to J. G. Siegel, 41/ the laser industry in the United States is expected to grow at a rate of approximately 16 to 20 per cent per year through 1990, at which time the sales of lasers including services will reach some $5,000 million.

Latest developments concerning the integration between robot mechanism and laser-based systems are presented, _inter alia_, in Lasers and Optics International. 42/

40/ B. Krauskopf, "Laser machining - no longer non-traditional". Manufacturing Engineering, October 1984.

41/ "Decision Resources". Arthur D. Little Inc., 1984.

42/ Lasers and Optics International. Monthly newsletter, vol. 1, No. 1, April 1984, published by IFS (Publications).

IX. APPROACH USED IN COLLECTING INFORMATION FOR THE STUDY

In accordance with the outline of the present study adopted by the Working Party on Engineering Industries and Automation at its second session, chapter VI is based on national contributions concerning the review of robot population by country, type of robot and area of application.

This chapter describes the approach to the above-mentioned task, i.e. the contents and form of the questionnaire on robot diffusion issued by the secretariat.

From the preceding chapters it can be seen that variations in data and information on robot population (total number of robots installed), annual production and international trade figures, and especially all forecasts, are caused mainly by two factors:

- Different definitions of robots accepted in various countries (manipulator as opposed to robot);

- Various approaches towards classification, namely when characterizing the nature of the robot control system.

Therefore, for the purpose of this study, when proposing a relatively simple and understandable classification which might serve in the near future for the collection of relevant information within the regular Annual Review of the Engineering Industries and Automation, (see chapter I) the following principles have been kept in mind:

(a) Indications of both the number of units (physical amount) and the value (in monetary terms) are necessary for defining the structure of production and international trade in robots;

(b) The description of the movable parts is a technical matter not necessarily needed for a general socio-economic analysis at the first stage of the Working Party's consideration in this field;

(c) An analysis of the areas of application, i.e. manufacturing, handling and assembly processes where the workpiece or tool is handled by a robot arm, may show the degree of automation of various processes. Among the most robotized technologies one will probably find spot and arc welding, paint spraying, moulding, loading and unloading of NC-machine tools and so on; and

(d) Diffusion of robots should be studied by branch and sector, either in the engineering industries themselves (automotive industry, electrical and electronic machinery, non-electrical machinery) or in other non-engineering sectors (plastics manufacturing, food industry, research and education etc.).

For chapter VI of the present study, i.e. the diffusion of robots in the ECE region, based on national contributions, two simple questionnaires were prepared (see annexes I and II) based on the following criteria:

Criterion 1: Type of robot

1.1. Programmable robot

Description: Computer (mini, micro)-controlled, off-line or on-line measuring device with servocontrol, usually continuous path control and error-correction. Wide use. Also called second-generation robot.

1.2. Robot with sensory perception

Description: Also called intelligent robot (based on principles of artificial intelligence) or third-generation robot. Equipped by optical and/or tactile sensors for recognition of colour, shape, weight, temperature etc. Self-teaching. Sophisticated software. Research in this field is still continuing.

Criterion 2: Operational mode

2.1. Workpiece gripped by robot

    2.1.1. Handling and transport operations
    2.1.2. Loading and unloading of machines
    2.1.3. Others

2.2. Tool handled by robot

    2.2.1. Joining of materials including welding
    2.2.2. Surface treatment including paint spraying
    2.2.3. Other metalworking
    2.2.4. Others

2.3. Assembly operations

Criterion 3: Application field

3.1. Engineering sectors

    3.1.1. Motor vehicles (ISIC 3843) 1/
    3.1.2. Electrical and electronic engineering (ISIC 383)
    3.1.3. Metal-working (non-electrical) engineering (ISIC 381, 382)
    3.1.4. Other engineering sectors (ISIC 38 excluding 381, 382, 383, 3843)

3.2. Other industrial sectors

    3.2.1. Ferrous and non-ferrous metals (ISIC 37)
    3.2.2. Plastic products (ISIC 3513, 3560)
    3.2.3. Other chemicals (ISIC 35 excluding 3513, 3560)
    3.2.4. Construction (ISIC 50)
    3.2.5. Textile industry (ISIC 32)
    3.2.6. Food industry (ISIC 31)
    3.2.7. Other industrial sectors (excluding services) (ISIC major
            divisions 2, 4, 3 excluding 31, 32, 35, 37)

1/ International Standard Industrial Classification, Revision 2.

3.3. <u>Non-industrial applications</u>

    3.3.1. Research and development
    3.3.2. Education and training
    3.3.3. Other non-industrial applications (agriculture, services etc.)

Two questionnaires were prepared (see annexes I and II) and distributed among ECE member countries:

    Questionnaire A: Interaction of criteria 1 and 2

    Questionnaire B: Interaction of criteria 1 and 3.

In both questionnaires, data were requested on: 2/

- Production in 1982

- Exports in 1982

- Imports in 1982

- Total robot population at the end of 1982.

As already mentioned, many government and non-governmental bodies, at both international and national levels, are undertaking special studies on the impact of new technologies, including robotics, on economic growth. To facilitate this task, it is essential that any feasible forecasting of various trends in this area should be based on reliable statistical data concerning production, use and international trade. In this respect, ECE activities in this field should be closely co-ordinated with those of other inter-governmental organizations such as OECD, EEC and CMEA. For example, the British Robot Association is now taking an initiative within the European Economic Community to co-ordinate statistical work concerning robot diffusion in the EEC region as from 1984. 3/ Similar activities might significantly help to improve future coverage of ECE industrial statistics, by including new automation-oriented engineering products in the regular statistics. Comments on the usefulness of the ECE efforts in assessing the diffusion of automation and its techno-economic aspects have already been made by several ECE member countries and such international organizations as the International Federation of Automatic Control (IFAC) 4/ (see also chapter X).

2/ Number of units and value in national currency.

3/ Information received from the Government of the United Kingdom.

4/ "UN/ECE Working Party on Engineering Industries and Automation".
IFAC Newsletter, No. 2, March 1983.

X. CLOSING REMARKS

One of the main current problems concerning economic growth is certainly that of finding ways to increase productivity. As described by Greenborough, 1/ "the tactical side of productivity means enhancing the effectiveness with which a given team of people, and a given range of equipment, achieve their purpose", "while in strategic terms, it is concerned with altering the number of people on the team and, a priori, independently, the range or complexity of the capital equipment they need to do the job". In other words, the level of productivity increases with higher output and lower costs per unit of production (labour, investments, energy and raw materials). As mentioned in chapter VIII, it is the potential slow-down of general productivity growth which forces the competent authorities to undertake technological change and introduce promising innovative programmes; however, the exact objectives may vary from country to country.

The experience gained from the installation of industrial robots during the last years (some 30,000 programmable robots were installed at the end of 1981, 39,000 at the end of 1982 and 50,000 at the end of 1983) might serve as a useful basis for a more general compilation of the over-all benefits and costs. As stated by J. Engelberger as cited in 2/ "... the industrial revolution was born and you're about to embark on another version of it. The third industrial revolution, if you want to call it that. And that says it will have a dramatic impact. If we succeed in sensory perception, you'll get people out of the mines. We don't have the technical answers yet, but coal mining shouldn't be done by people. You'll end up with a shorter working week and a better life style. You will be wealthy enough to support it, certainly as you're now supporting all those people on redundancies ...".

As mentioned above (see chapter VIII) robot installation costs comprise:

- The purchase price of the robot installation;

- Maintenance costs (both hardware and software);

- Operation costs;

- Depreciation; and

- Personnel costs.

The purchase price still represents at least 50 per cent of total costs: the special tooling, tailored accessories and peripherals are still very expensive as opposed, however, to the decreasing costs of microelectronic-based controls. Research and development activities are therefore oriented towards standardized and modular solutions.

1/ J.H. Greenborough, Introduction to "Productivity Measurement", edited by D. Bailey and T. Hubert. Gower Publishing Co. Ltd. (for the British Council of Productivity Associations), 1980.

2/ J. Bell "We'll have robots like ladies have hats". New Scientist, 24 February 1983.

On the other hand, the main benefits of using robots are:

- Increased output (throughput, productivity, etc.);

- Labour-cost savings;

- Higher quality of the manufacturing process of the products themselves; and

- Other savings (raw materials, energy etc.).

According to most of the available case studies on robot profitability, the main economic benefit lies in the area of labour saving, if the robot is used for two or three shifts per day. The pay-off time for successful implementations is usually between two and three years.

During the past few years robots have become a reality for users in many countries. Their application is no longer a vision of tomorrow, but a practical way of solving problems of labour productivity, increasing quantity and speed of mass production, meeting higher qualitative demands, replacing humans in dangerous work and so on. The availability of sophisticated, miniaturized and relatively cheap hardware, due to the unexpected success of microelectronics, seems to be the most important factor influencing the wide diffusion of robots. Serious technical problems, however, remain to be solved, such as the need for

- Continuous innovation of robot mechanisms to permit higher operational
  speed, manipulation of heavier loads, increased accuracy, reliability
  and safety, easy maintenance, etc.;

- Design and production of suitable vision, tactile, acoustic and other
  sensors located on or close to the robot's end effectors; and

- Rapid development of appropriate software tools, both standardized
  packages for universal tasks and interface with the whole system
  (other machines, man/machine interfaces, integration with information
  functions, etc.) and special tailored application programs for various
  manufacturing processes and operational modes.

As in other industrial automation areas such as the introduction of NC-machine tools, computer-aided design and manufacturing, and flexible manufacturing systems, special attention should be given to:

- Methods of economic justification for the introduction of robots into
  manufacturing and assembly lines and assessment of their direct and
  indirect economic effects; and

- Training and education of personnel, both specialists (electronic,
  mechanical and production engineering) and supervisors (skilled
  operators and maintenance personnel).

In this field, international co-operation within ECE might be useful, bearing in mind such activities as exchange of information, implementation experience, standardization and promotion of technical and scientific co-operation and trade. As mentioned above, a first step in this direction could be the regular collection by the ECE secretariat of relevant data and national views concerning the assessment of the diffusion of robots by type, operational mode and area of application.

Technological innovation appears to have no limits. Recent developments in various areas of industrial automation have created the necessary conditions for total, unmanned automation of manufacturing, servicing, etc., pointing to the necessity of an integrated approach towards the introduction of robots and flexible automation. 3/ The technical, economic and social barriers to a broader application could be overcome only if management were able to ensure the necessary adaptability of organizational and economic structures, preparation of personnel and other "environmental" consequences.

3/ H.D. Haustein and H. Maier, "The diffusion of flexible automation and robots". IIASA Laxenburg, November 1981.

Figure 1. Interactions between information-processing and production-control functions

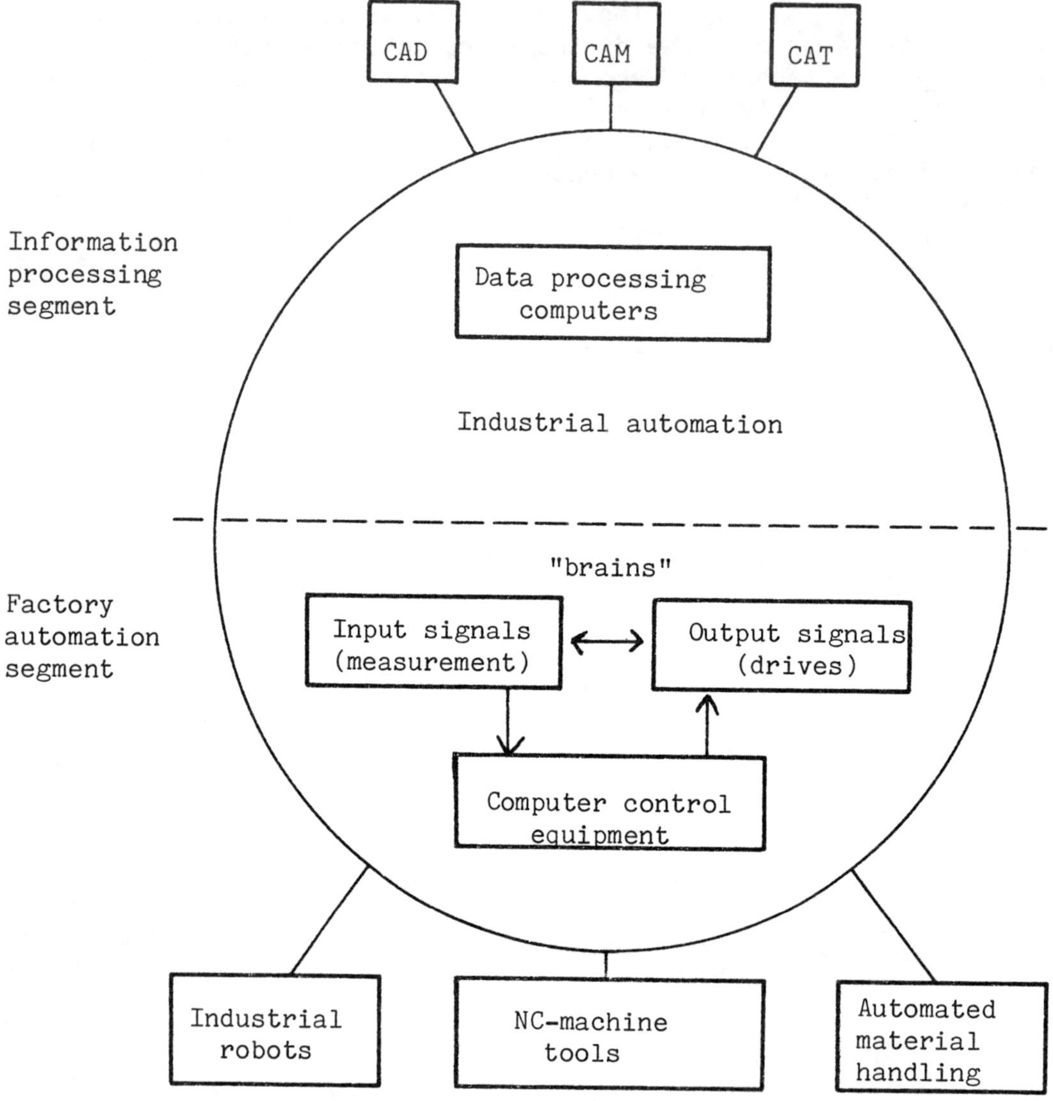

Source: C.A. Hodson "Computers in Manufacturing"; Science, vol. 215, February 1982.

Figure 2. Indication of functional relation between flexibility and productivity for different types of equipment

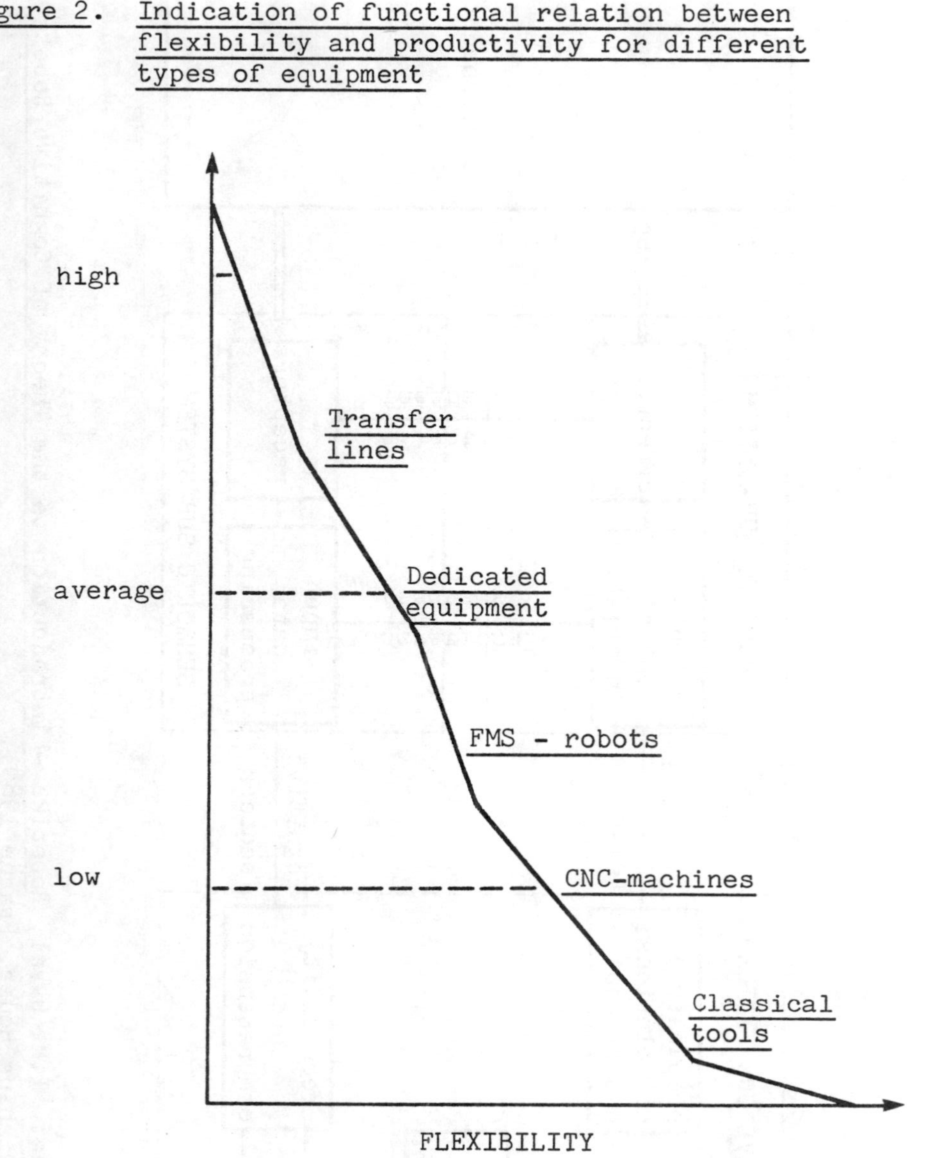

Source: T. Palazzo, "Maximizing Production – The Answer: FMS"; Proceedings of the Conference, 5th NC Industrial Automation and Robot Exhibition, Milan, 1-5 March 1982.

Figure 3. Decomposition of the robot system into sub-systems

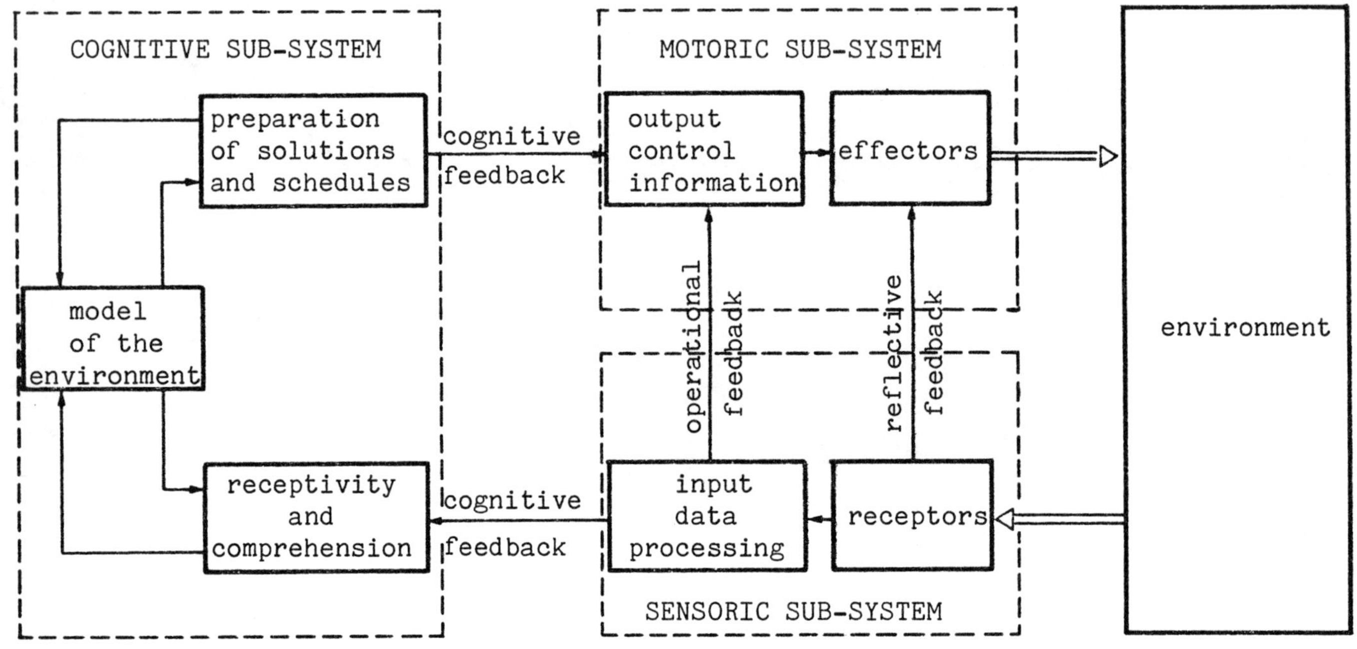

Source: I.M. Havel, Robotika - Introduction to the Theory of Cognitive Robots (SNTL Publishing House, Prague 1980).

Figure 4. Distribution of industrial robots by
number of articulations

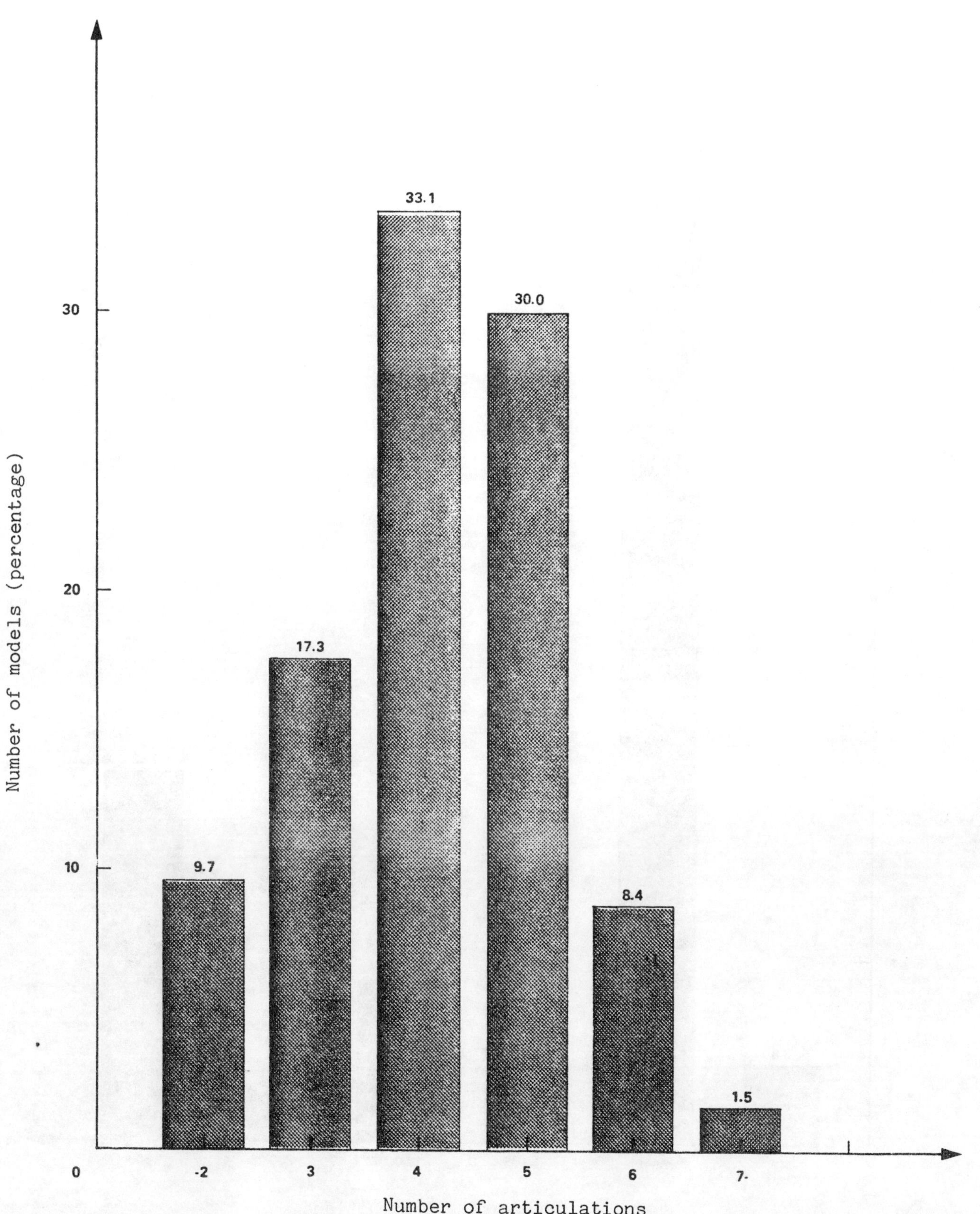

Source: Y.G. Kozyrev, "Perspectives of Development and Use of Industrial Robots" (submitted by the Government of the Union of Soviet Socialist Republics).

Figure 5. Distribution of industrial robots by number of programming instructions

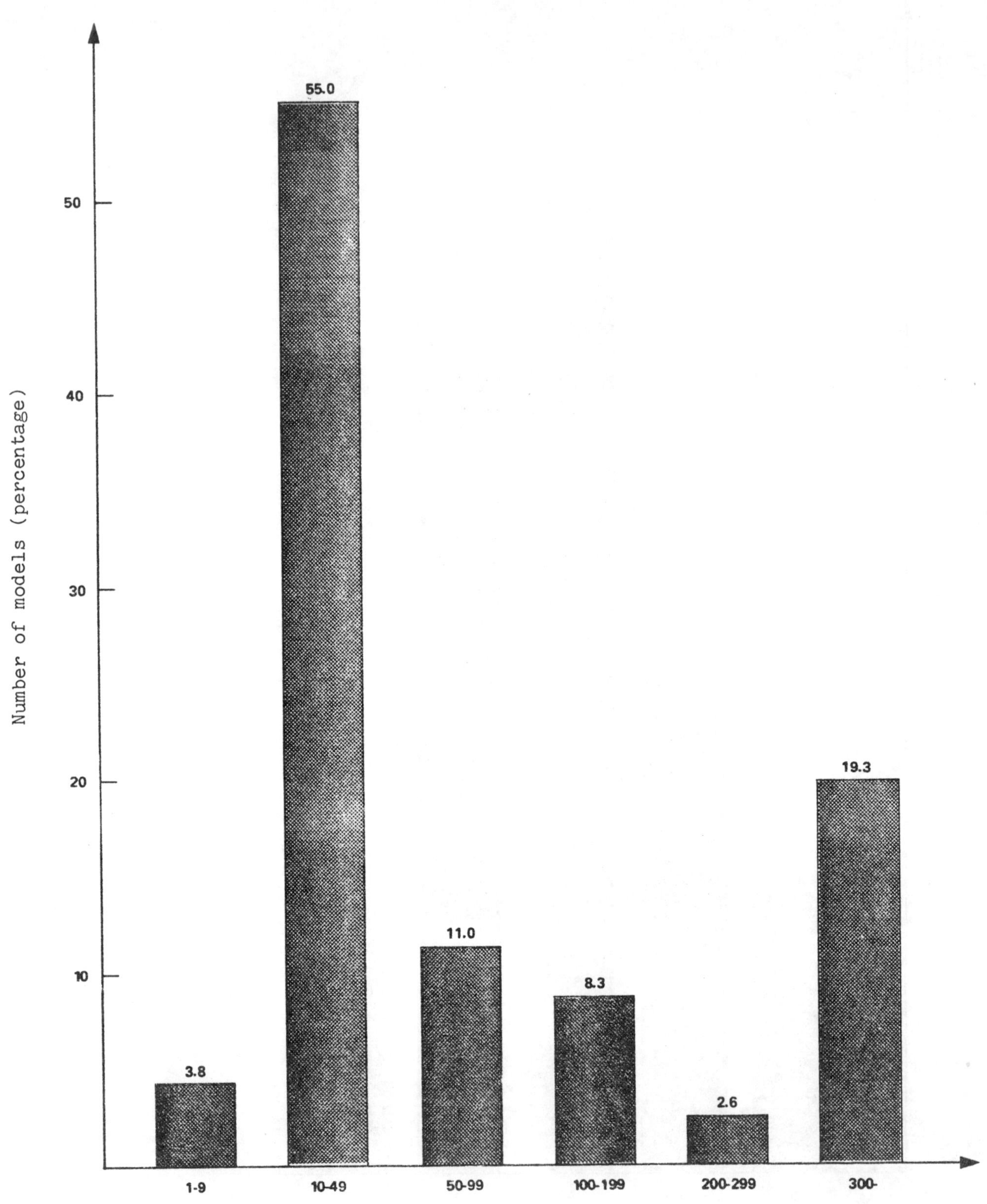

Source: Y.G. Kozyrev, "Perspectives of Development and Use of Industrial Robots" (submitted by the Government of the Union of Soviet Socialist Republics).

Figure 6. Distribution of industrial robots by number of
input/output interface channels

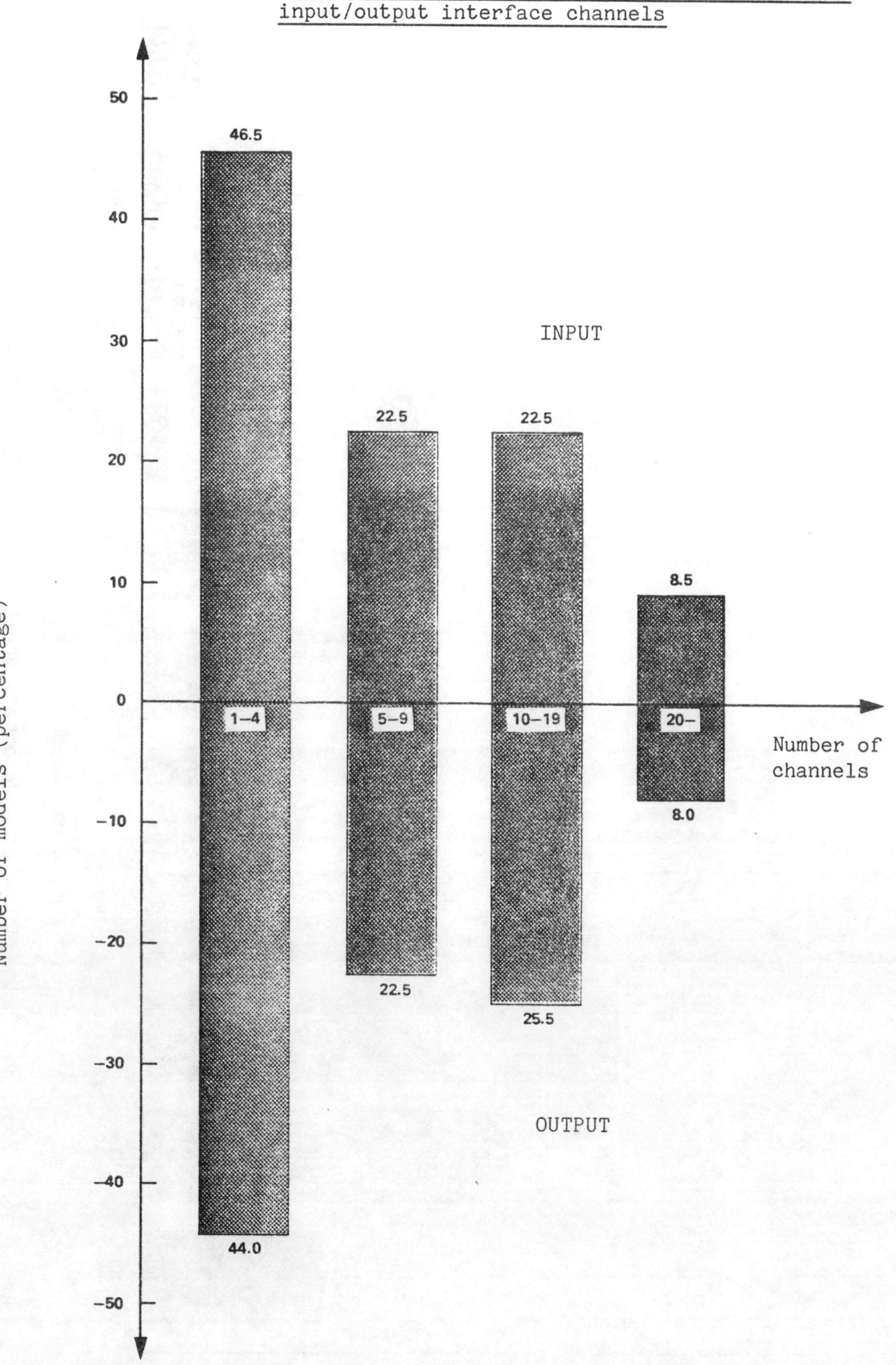

Source: Y.G. Kozyrev, "Perspectives of Development and Use of Industrial
Robots" (submitted by the Government of the Union of Soviet Socialist Republics).

Figure 7. Distribution of industrial robots by load capacity

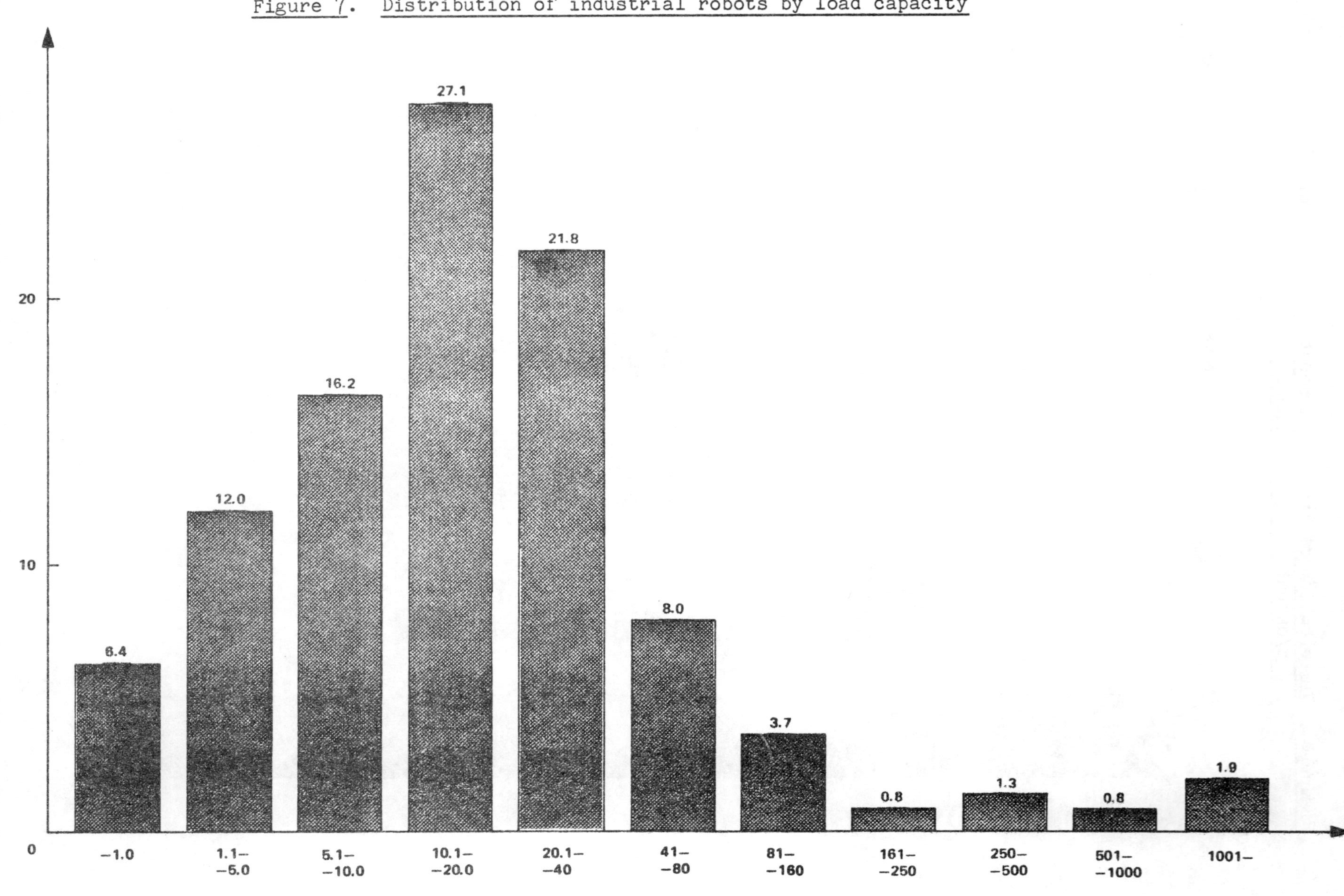

Source: Y.G. Kozyrev, "Perspectives of Development and Use of Industrial Robots" (submitted by the Government of the Union of Soviet Socialist Republics).

Figure 8. Distribution of industrial robots
by working (operating) space

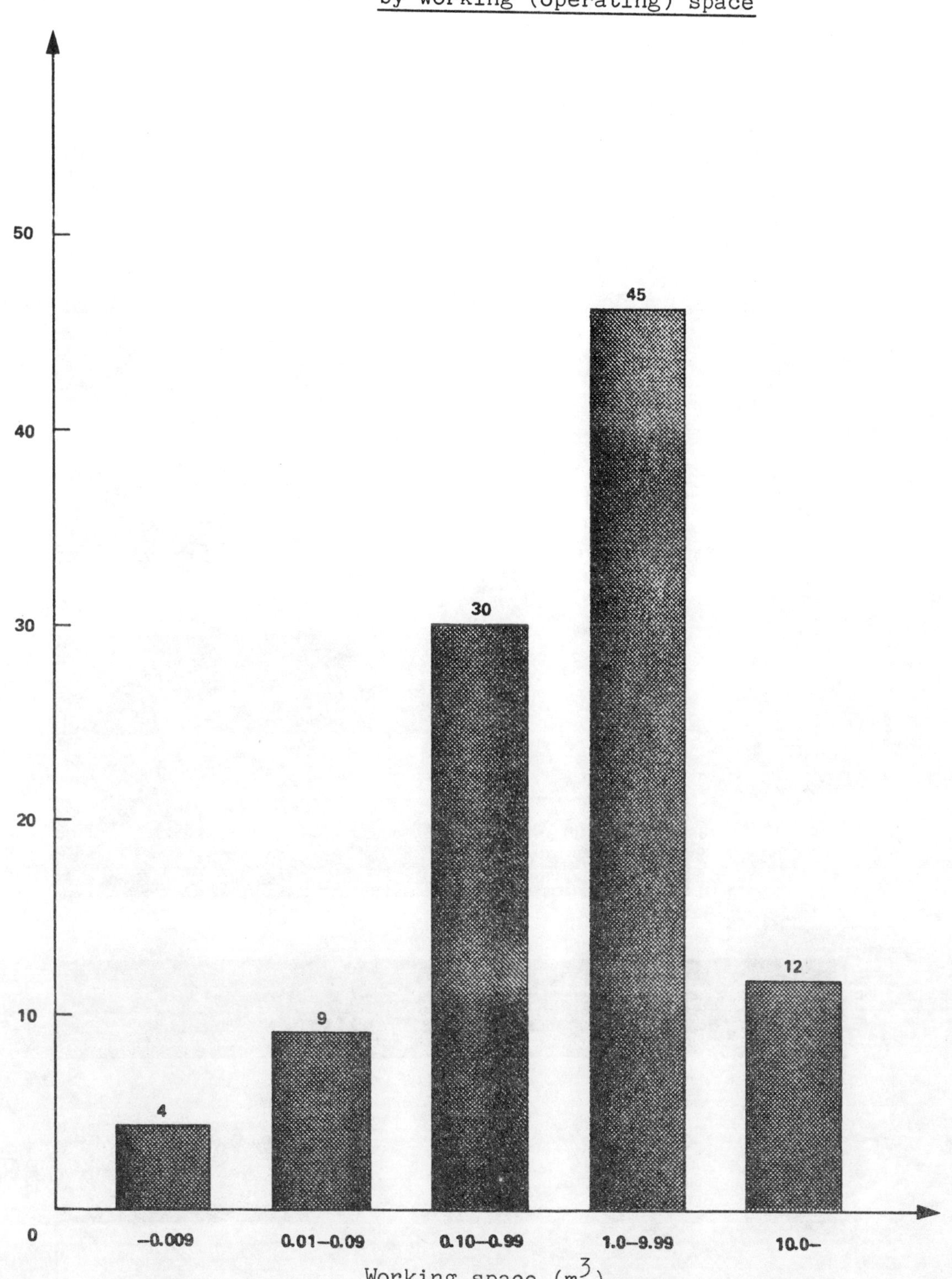

Source: Y.G. Kozyrev, "Perspectives of Development and Use of Industrial
Robots" (submitted by the Government of the Union of Soviet Socialist Republics).

Figure 9. Distribution of industrial robots by positioning error

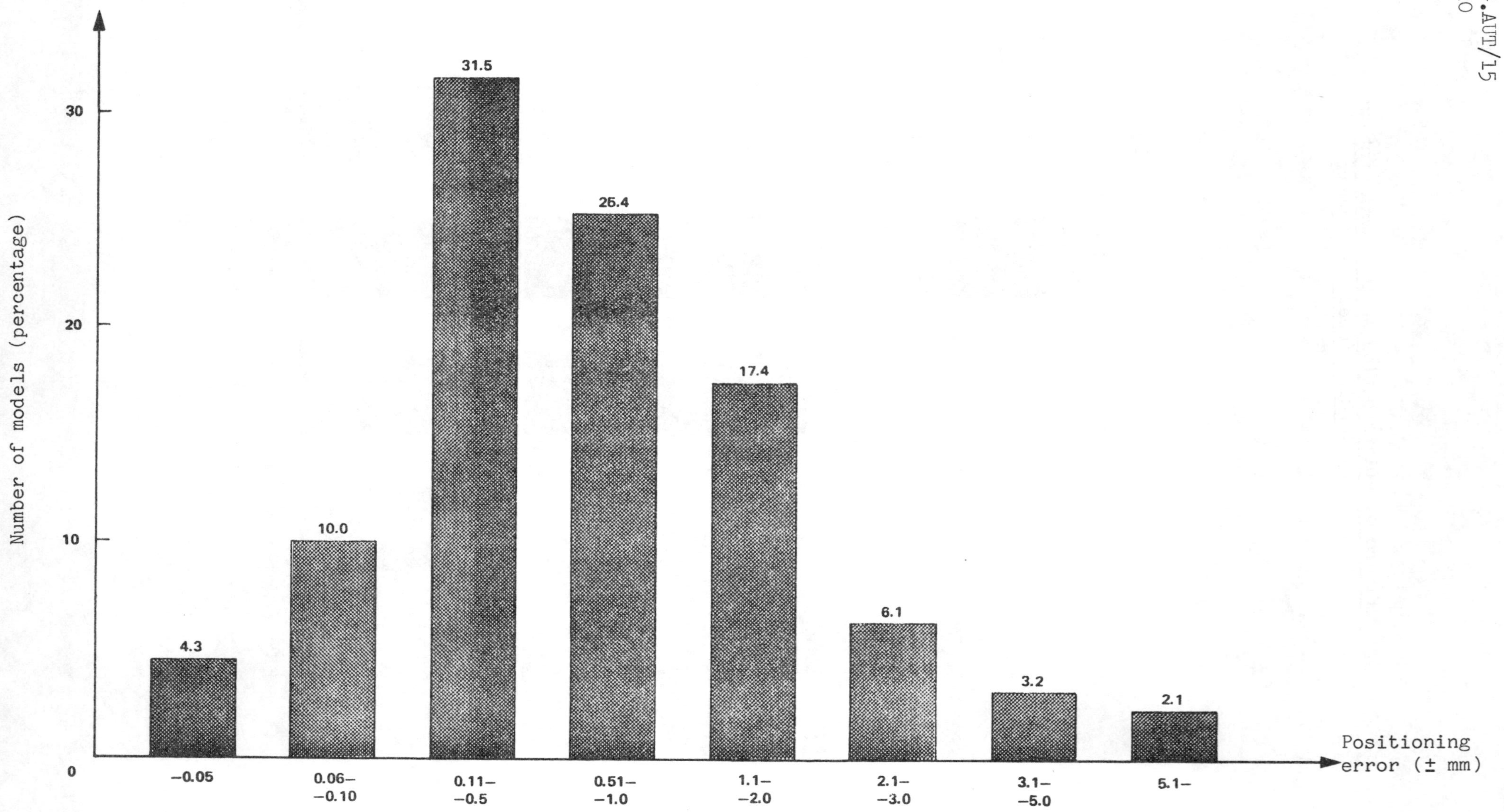

Source: Y.G. Kozyrev, "Perspectives of Development and Use of Industrial Robots" (submitted by the Government of the Union of Soviet Socialist Republics).

Figure 10.   Robot-system interfaces

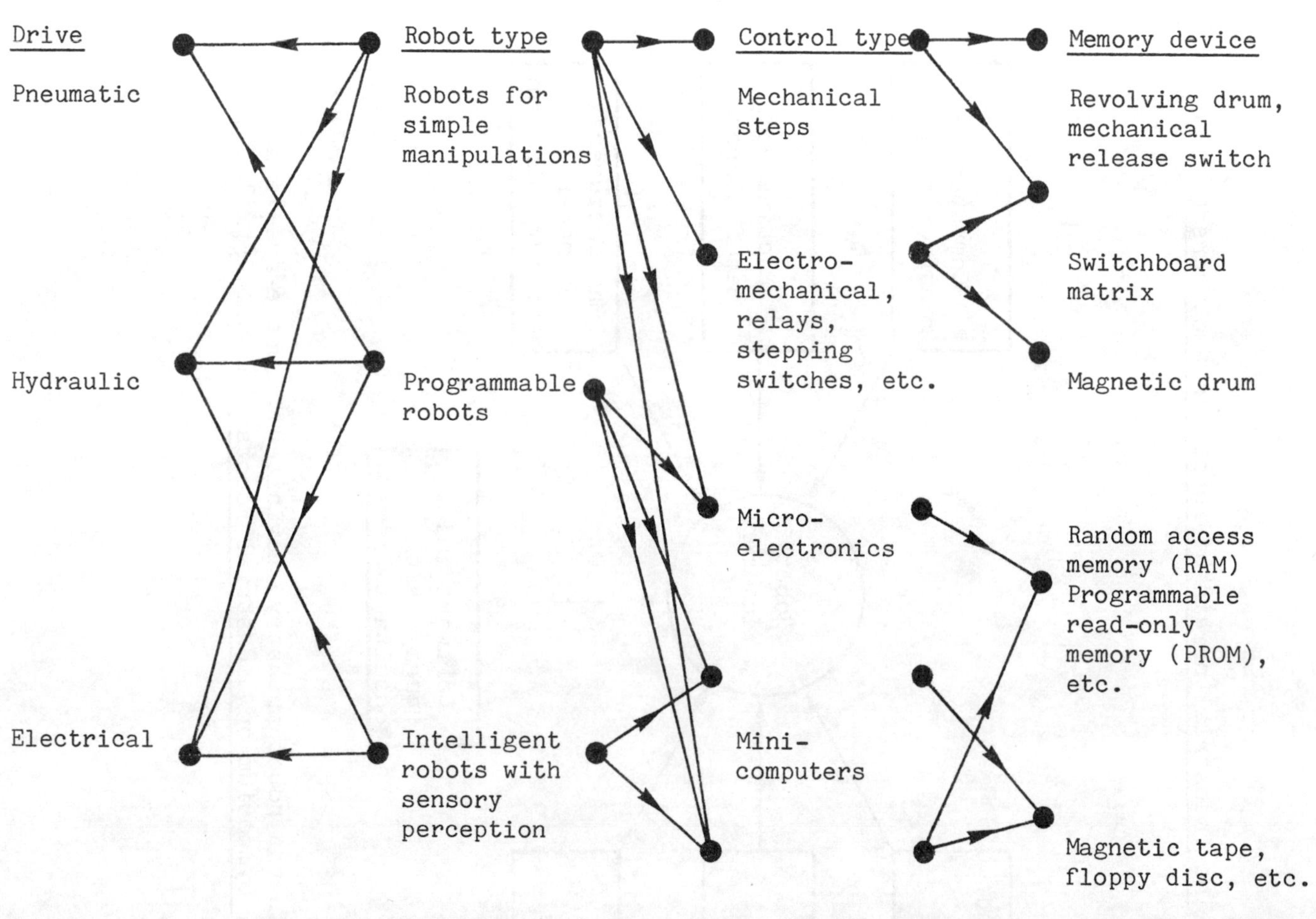

Source: Mackintosh Consultants Co. Ltd., Luton 1979.

Figure 11.   The industrial origin of robot manufacturers

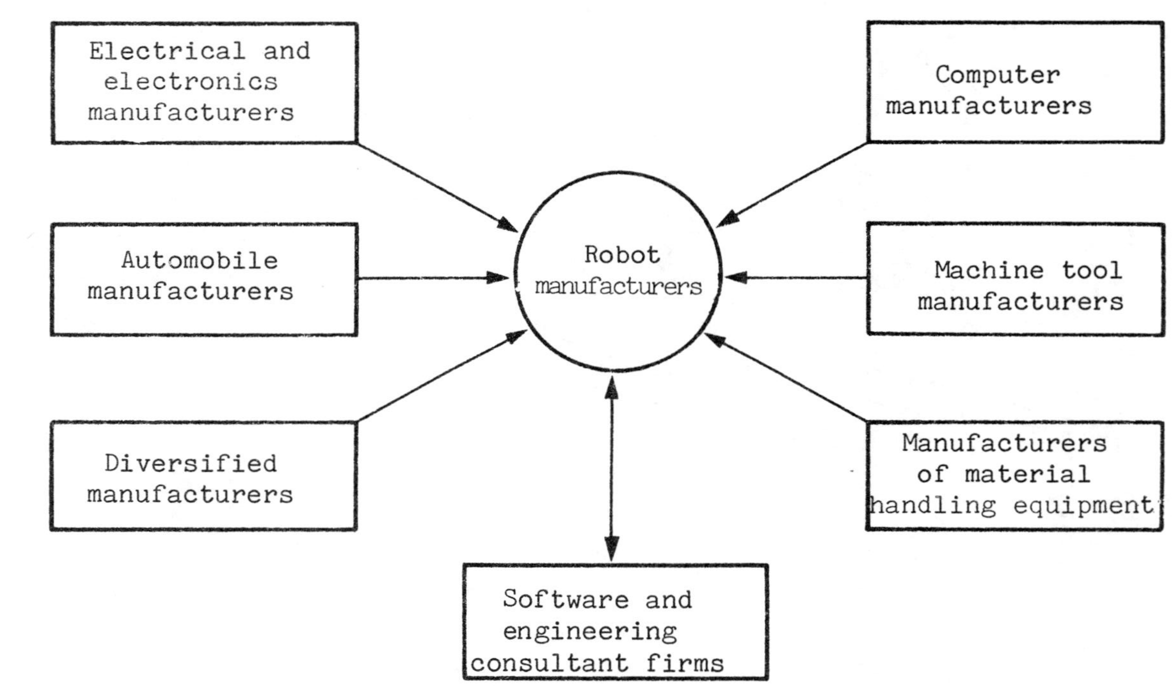

Source: L. Conigliaro, Trends in the robot industry (revisited): where are we now? Proceedings of the 13th International Symposium on Industrial Robots, Chicago, April 1983.

Figure 12. Production of automated manufacturing equipment
and systems (AMES) - Comparative strategies in
the United States, Western Europe and Japan

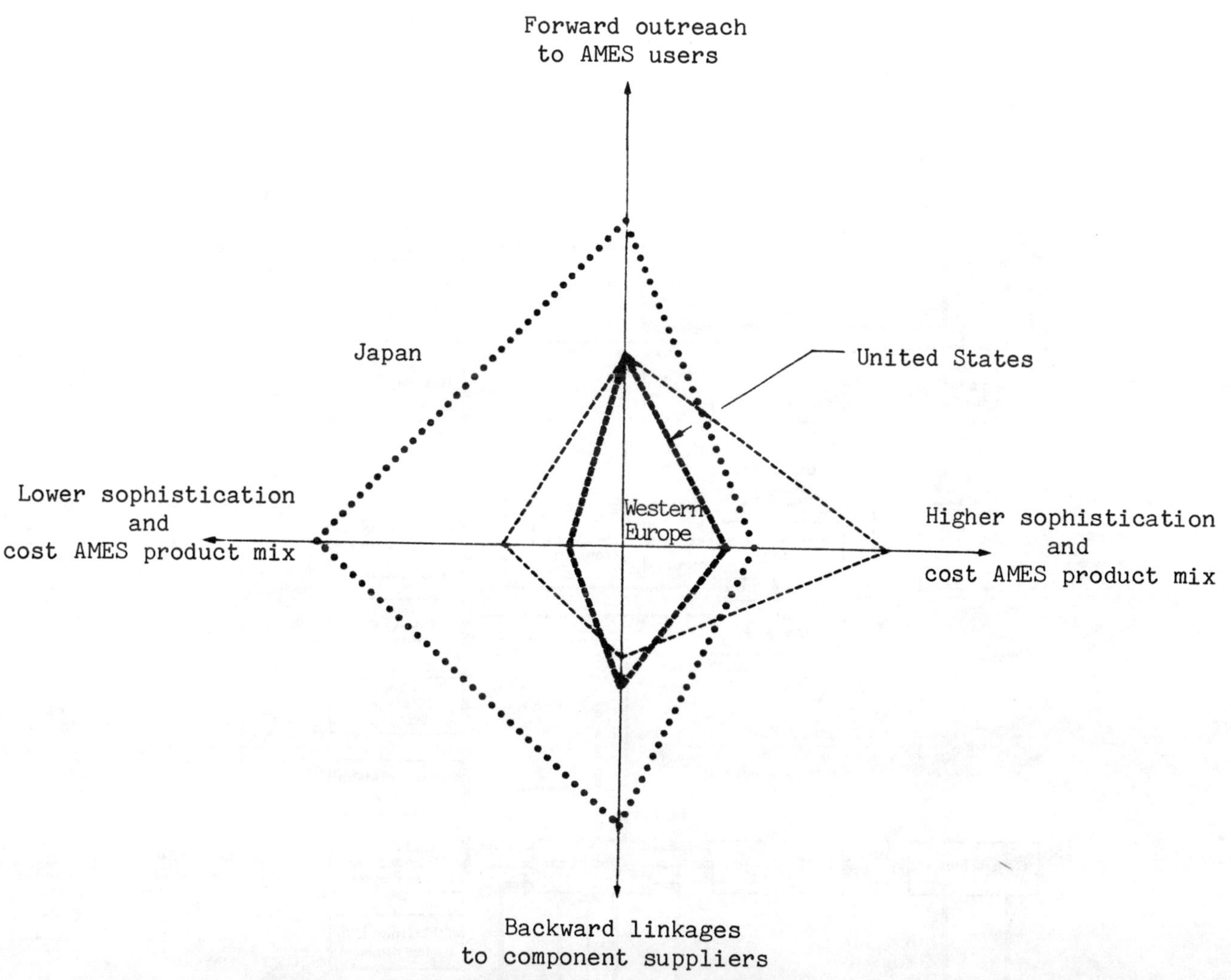

Source: J. Baransow, Robots in manufacturing - Key to international competitiveness.
Lomond Publications Inc. Maryland, United States 1983.

FIGURE 13a

International co-operation in the field of industrial robots

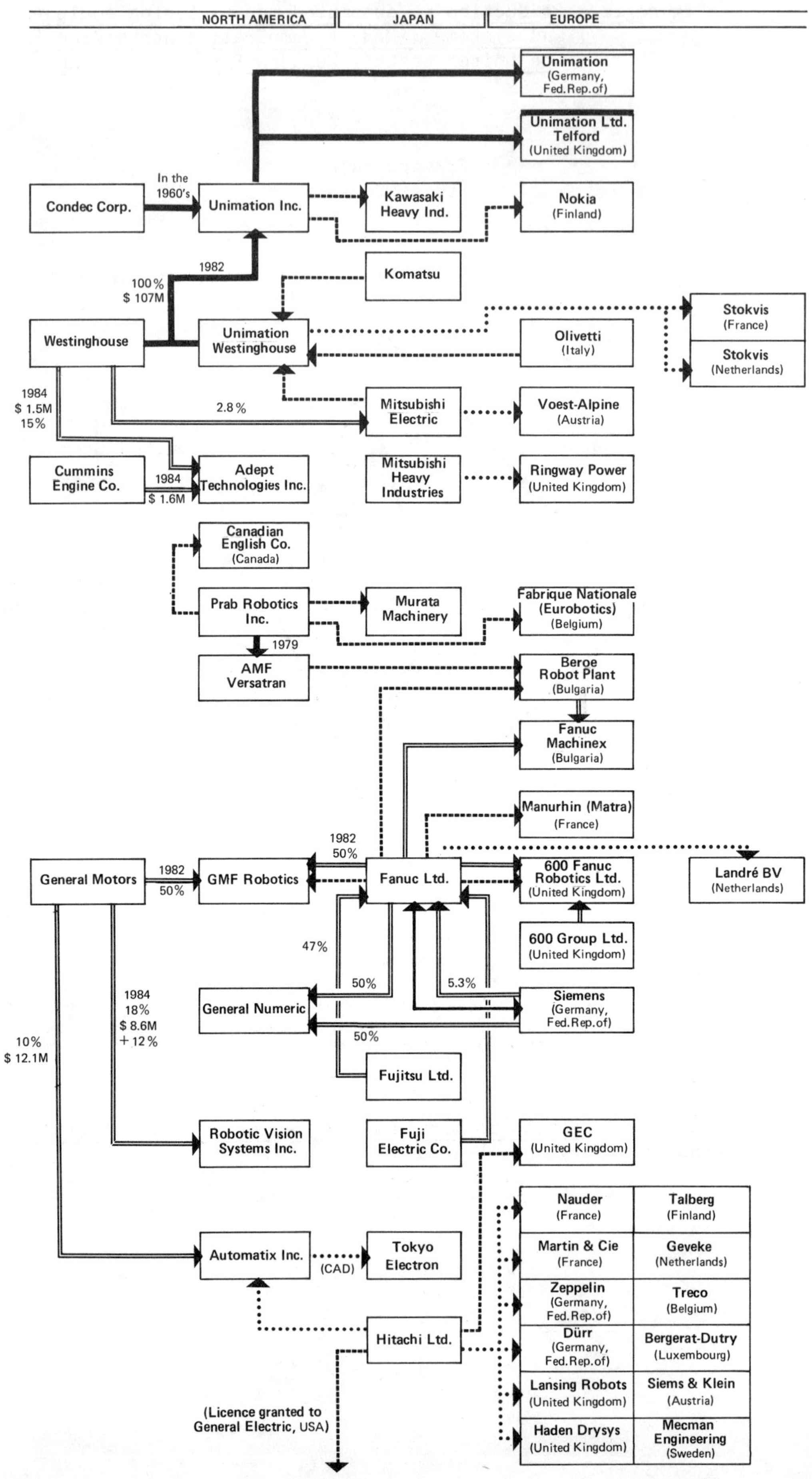

FIGURE 13b
International co-operation in the field of industrial robots

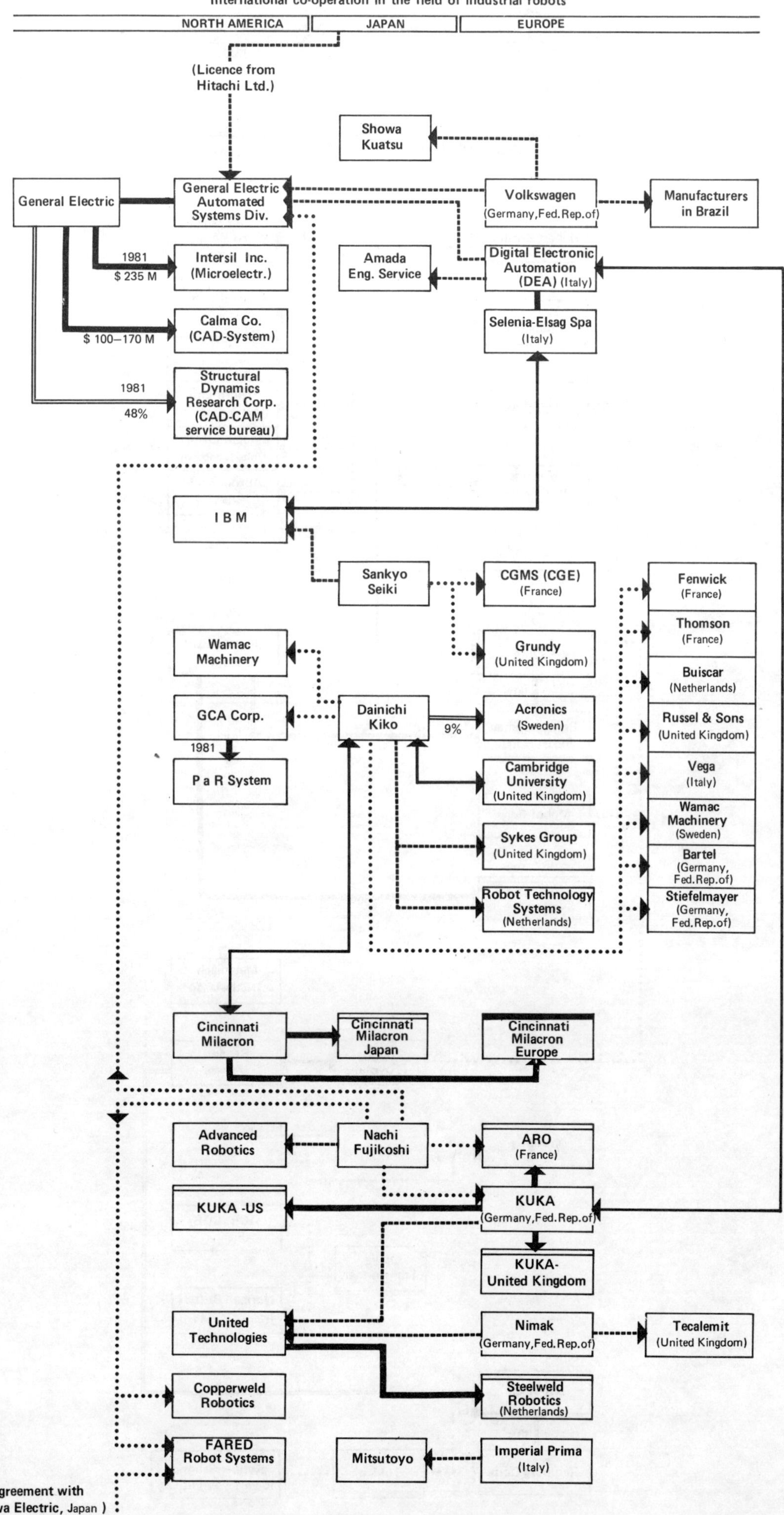

FIGURE 13c
International co-operation in the field of industrial robots

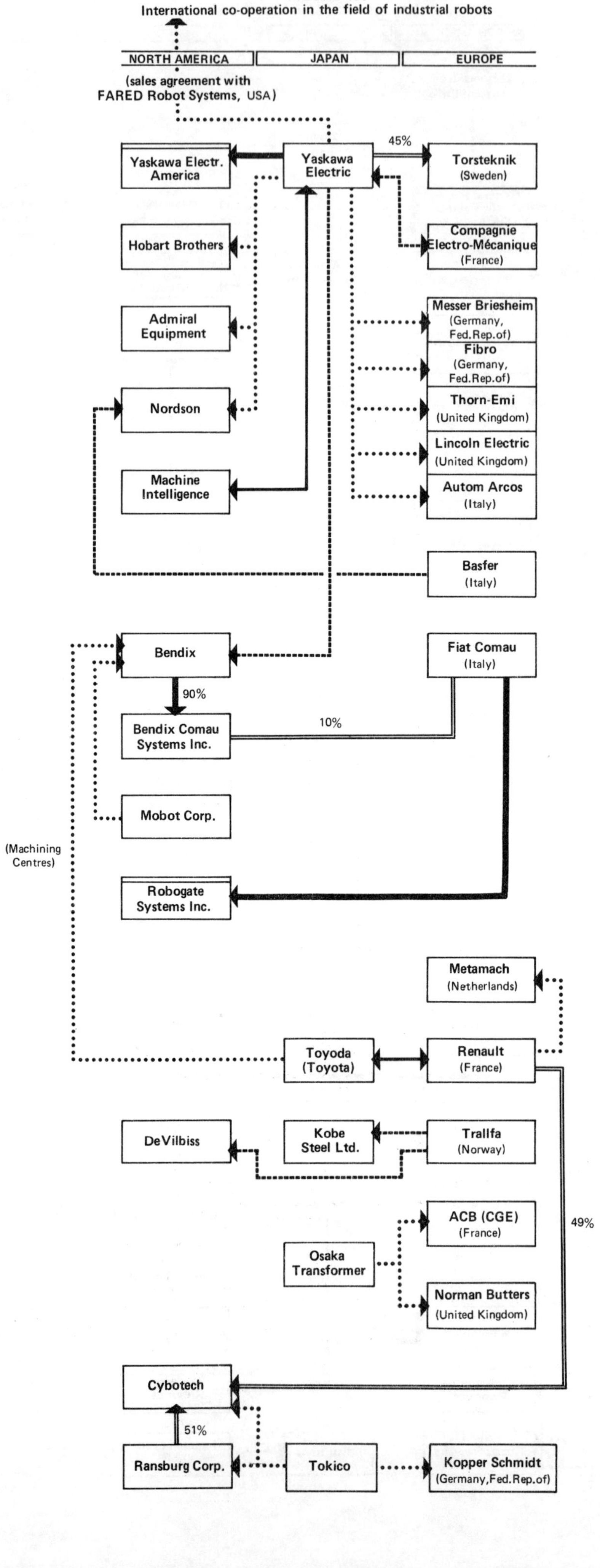

FIGURE 13d
International co-operation in the field of industrial robots

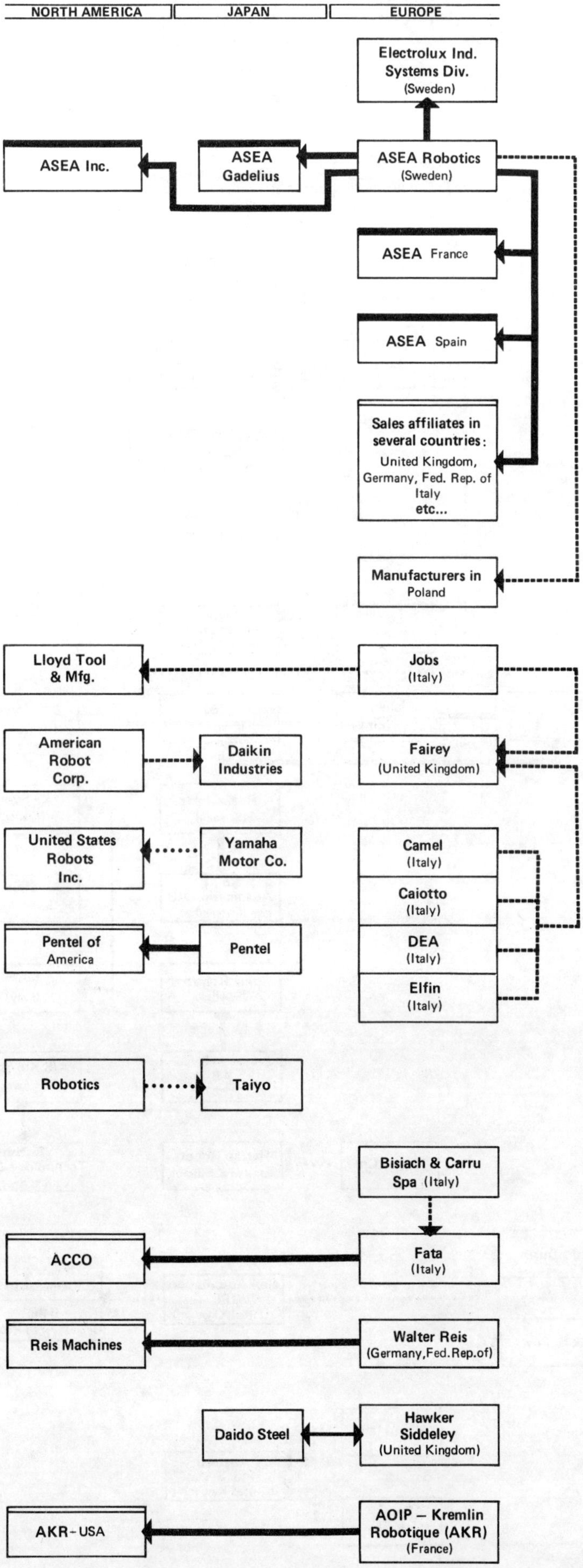

FIGURE 13e
International co-operation in the field of industrial robots

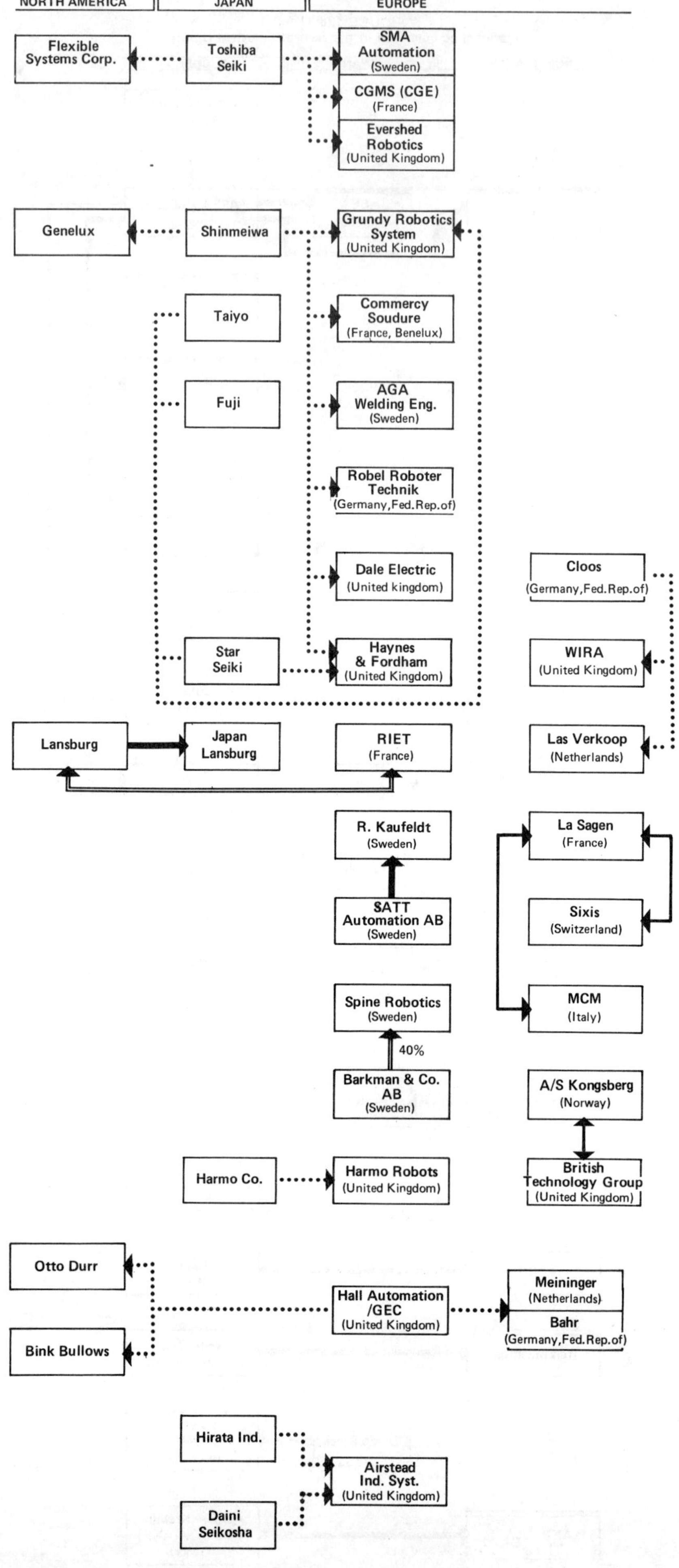

FIGURE 13f
International co-operation in the field of industrial robots

KEYS TO SYMBOLS

Parent company; industrial robot subsidiary company; industrial robot division within parent company.

Foreign industrial robot manufacturing subsidiary (hardware, software, peripherals).

Foreign industrial robot sales subsidiary

Subsidiary (75%—100% shareholding) (arrow pointing from parent to subsidiary company).

Joint venture (parent company holding less than 75% of shares).

Licences for manufacturing and sales

Mutual licensing of the industrial robots of the two companies

Marketing and sales agreement

Mutual marketing and sales agreement

Joint venture, without shareholding, in research and development, technology exchange, manufacturing.

Source: ECE secretariat (see chapter VII)

Figure 14. Indication of cost efficiency by
manufacturing mode

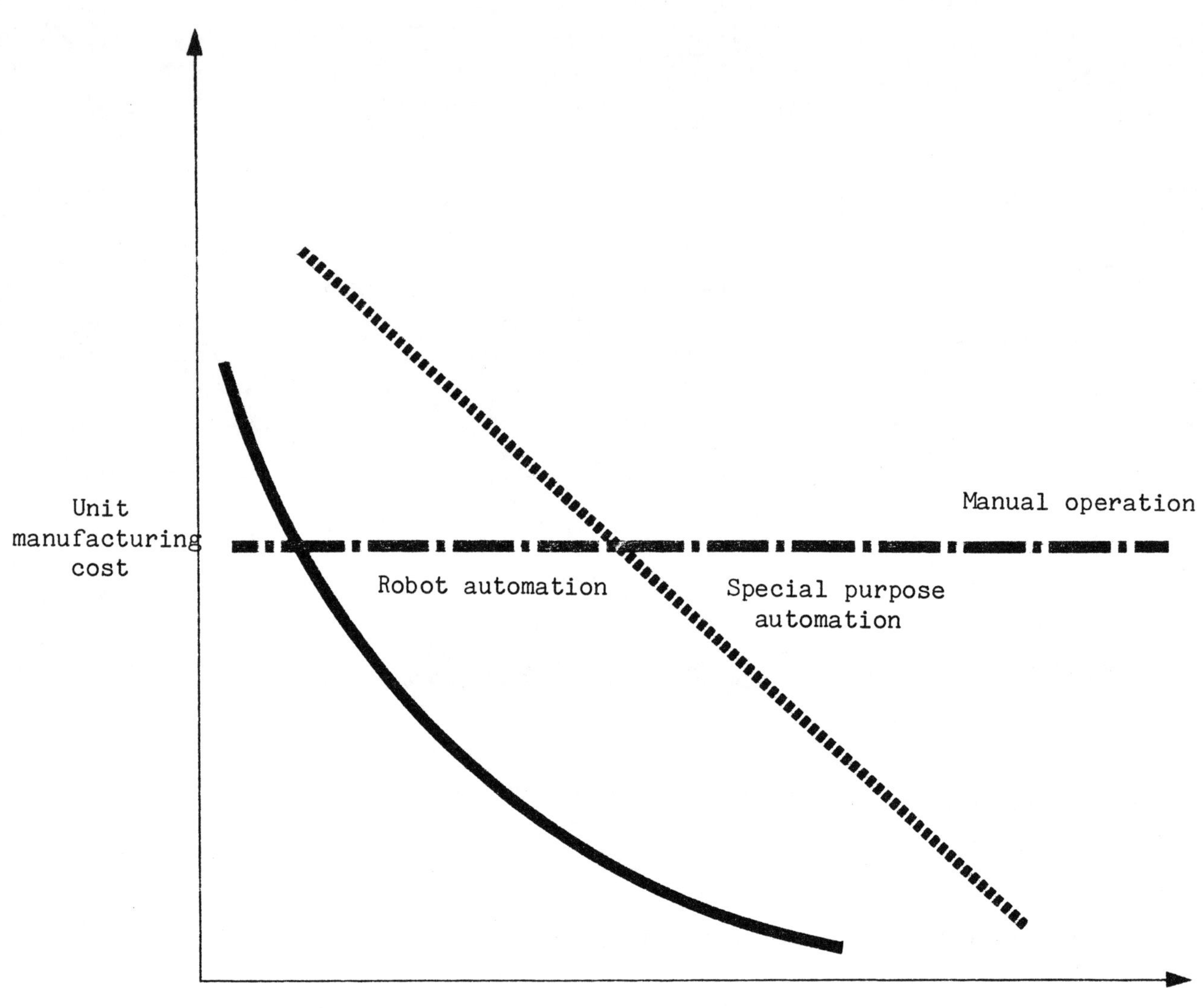

Source: R. Zermeno-Gonzalez, A Conceptual Framework for the Study of the Adoption of Robots in Manufacturing Industry, International Institute for Applied Systems Analysis (IIASA); Laxenburg, December 1979.

Figure 15. Typical structure of a machine vision process

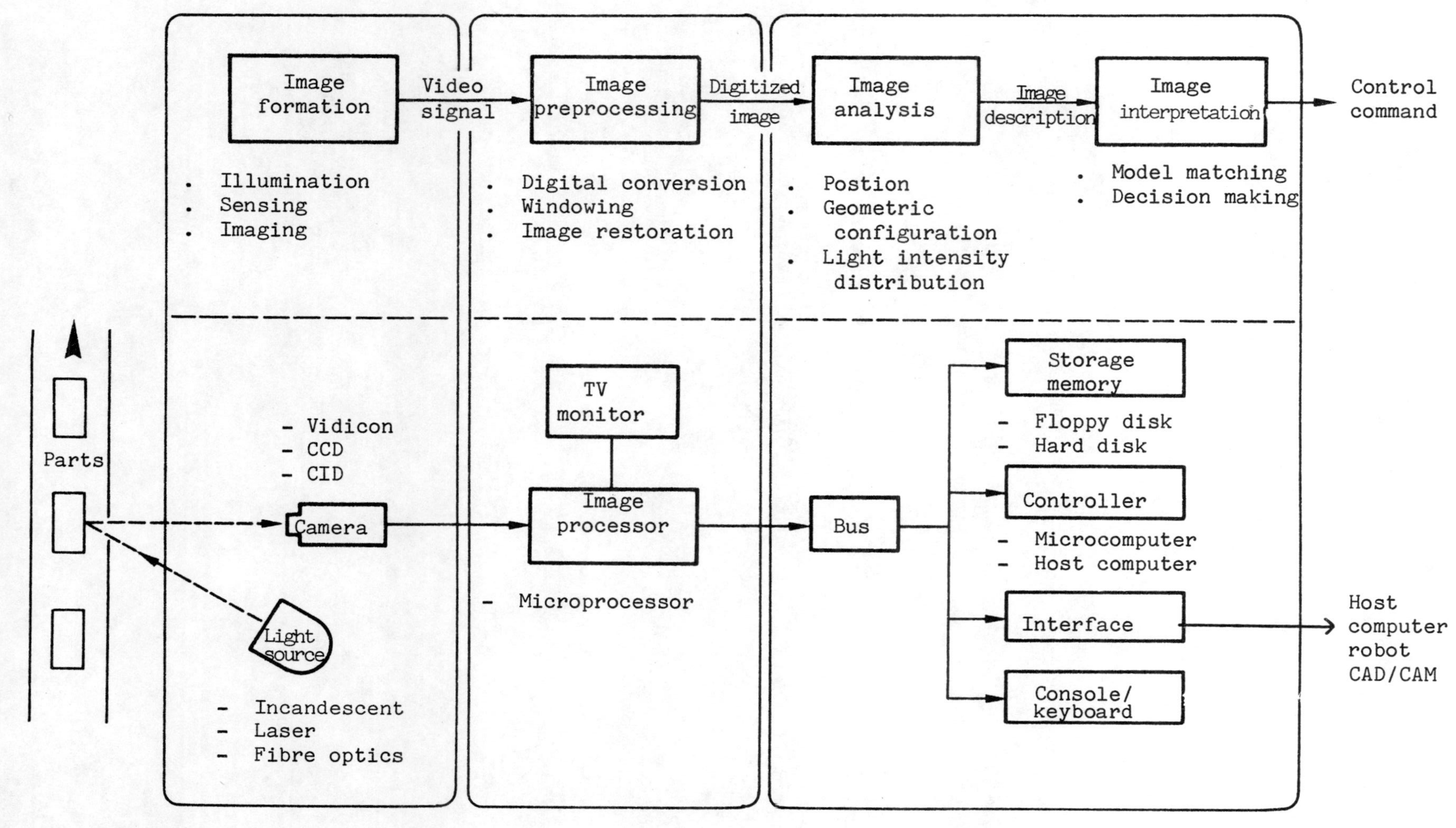

Source: G. Schaffer, "Machine vision a sense for computer integrated manufacturing" Special report 767, American Machinist, June 1984.

<u>Annex I</u>

QUESTIONNAIRE: DISTRIBUTION OF ROBOTS BY OPERATIONAL MODE

Country:

National currency:

Exchange rate:
(US dollars
per unit of national
currency)

Operational mode	Programmable robots								Robots with sensory perception							
	1982			Total robot population a/	1982			Total robot population a/	1982			Total robot population a/	1982			Total robot population a/
	Production	Exports	Imports		Production	Exports	Imports		Production	Exports	Imports		Production	Exports	Imports	
	Number of units				Value (thousands) National currency				Number of units				Value (thousands) National currency			
Handling and transport operations																
Loading and unloading of machines																
Others (workpiece gripped)																
WORKPIECE GRIPPED BY ROBOT SUB-TOTAL																
Joining including welding																
Surface treatment including paint spraying																
Other metal-working																
Others (tool handled)																
TOOL HANDLED BY ROBOT SUB-TOTAL																
Assembly operations																
TOTAL																

a/ **As** of end 1982.

Annex II

QUESTIONNAIRE B: DISTRIBUTION OF ROBOTS BY FIELD OF APPLICATION

Country:

National currency:

Exchange rate:
(US dollars per
unit of national
currency)

Application field (sector, branch, etc.)	International Standard Industrial classification (ISIC,Rev.2)	Programmable robots								Robots with sensory perception							
		1982			a/	1982			a/	1982			a/	1982			a/
		Production	Exports	Imports	Total robot population	Production	Exports	Imports	Total robot population	Production	Exports	Imports	Total robot population	Production	Exports	Imports	Total robot population
		Number of units				Value (thousands) national currency				Number of units				Value (thousands) national currency			
Motor vehicles	3843																
Electrical and electronic engineering	383																
Metal-working (non-electrical) engineering	381, 382																
Other engineering	38 excluding 381, 382, 383, 3843																
ENGINEERING	SUB-TOTAL																
Ferrous and non-ferrous metals	37																
Plastic products	3513, 3560																
Other chemicals	35, excluding 3513, 3560																
Construction	50																
Textile industry	32																
Food industry	31																
Other industrial excluding services	Major divisions 2, 4, 3 excluding 31, 32, 35, 37																
OTHER INDUSTRIAL EXCLUDING ENGINEERING	SUB-TOTAL																
Research and development																	
Education and training																	
Other non-industrial (agriculture, services, etc.)																	
NON-INDUSTRIAL	SUB-TOTAL																
TOTAL																	

a/ As of end 1982.

<u>Annex III</u>

QUESTIONNAIRE FOR THE ANNUAL REVIEW OF THE
ENGINEERING INDUSTRIES AND AUTOMATION
(Revised)

I. Statistical Part

5. <u>Special statistics</u>

(Industrial Robots, 1983-1984)

	Units						Value (current prices, millions of national currency)					
	Production		Imports		Exports		Production		Imports		Exports	
	1983	1984	1983	1984	1983	1984	1983	1984	1983	1984	1983	1984
Industrial robots <u>a</u>/ TOTAL												
<u>of which:</u> Handling and Loading robots <u>b</u>/												
Processing robots <u>c</u>/												
Assembly robots												

    <u>a</u>/ Definition of industrial robots as proposed by the International Organization for Standardization (ISO).

    <u>b</u>/ Used for applications where the workpiece is gripped by the robot.

    <u>c</u>/ Used for applications where the tool is handled by the robot (e.g. joining including welding, surface treatment including paint spraying, other metal-working, etc.).

<u>Symbols to be employed</u>

    -    =   Magnitude zero

    0.0 =   Magnitude not zero, but less than half of unit employed

    ... =   Not available

    x   =   Provisional or estimated figure

    r   =   Revised

<u>Annex IV</u>

INTERNATIONAL ORGANIZATION FOR STANDARDIZATION (ISO)
Glossary of terms in industrial robotics (draft)

ACCURACY

A qualitative assessment of freedom from error or of the degree of conformity to
a desired value, a high assessment corresponding to a small error.

ACTUATOR

A power mechanism used to effect motions of the robot.

ADAPTIVE CONTROL

A method of continuous and automatic adjustment of control parameters in response
to measured process variables aimed at achieving a better performance.

APPLICATION WORKING SPACE

The volume of space needed by the end effector to perform a specified application.

ARM

An interconnected set of links and powered joints supporting and moving an end
effector

ARTIFICIAL INTELLIGENCE

The ability of an industrial robot to perform functions such as reasoning,
planning, problem solving, pattern recognition, perception, cognition,
understanding and learning.

AXIS DEFINITION

A direction in which a part of a robot can move in a linear or rotary mode.
The number of axes normally corresponds to the number of guided and mutually
independently driven links.

BASE

A pedestal to which is secured the origin of the first member of the articulated
structure which makes up the robot; the other end of this structure is the
coupling device.

CONTINUOUS PATH CONTROL

See annex V.

DEGREE OF FREEDOM

One of a limited number of ways in which a dynamic system may change, each way
being expressed by an independent variable. The number of degrees of freedom of
a robot normally relates to its ability to position and orient its end effector.
In this case its theoretical maximum value is six. Because of a possible
confusion with axes, it is advisable to avoid using this word for describing an
industrial robot.

DISTRIBUTED JOINT

An assembly between two rigid members enabling one to rotate or translate in
relation to the other in such a way that the rotation or translation is
distributed along the assembly.

DUTY CYCLE

The fraction of time during which a device of system is active or at full power.

DYNAMIC OR PATH ACCURACY

Degree of conformance to the fixed value as the relevant variables change with
time.

EFFECTOR COUPLING DEVICE (ECD)

A part securing the end effector to the end of the robot arm.

END EFFECTOR

A device (**gripper, tool,** sensor, measuring instrument etc.) secured to the wrist
by means of the effector coupling device and which performs the robot task.

FLEXIBLE

1.   Pliable or capable of bending. In robot mechanisms this may be due to joints,
links, or transmission elements. Flexibility allows the endpoint of the robot to
sag or deflect under the load and to vibrate as a result of acceleration or
deceleration.

2.   Multipurpose; adaptable; capable of being redirected, retrained or used
for a new purpose. Refers to the reprogrammability or multi-task capability of
robots.

GRIPPER

An end effector specialized in grasping and holding.

HAND

A specialized gripper consisting of jaws capable of grasping.

JOINT SPACE

The space defined by vector components of the angular or translational
displacement of each joint of a multi-axis linkage relative to the reference
displacement for each such joint.

LOAD CAPACITY AND PAYLOAD

<u>Nominal load or load capacity:</u>

The maximum load that can be applied to the coupling device without degradation
of any performance specification including speed, traverse time, working range
and accuracy. Nominal load (load capacity) is the same of tool load and net load.

<u>Tool load:</u>

The load of the end effector and any additional user-added equipment mounted on the coupling device.

<u>Net load or payload:</u>

The maximum load which in addition to the tool load can be applied to the coupling device without degradation of any performance specification including speed, traverse time, working range and accuracy.

<u>Maximum net load:</u>

The maximum load which in addition to the tool load can be handled without risking any permanent damage or failure of the robot mechanism. Performance specifications are necessarily restricted.

<u>Maximum load or limit load:</u>

The maximum load which can be handled without risking any permanent damage or failure of the robot mechanism. Performance specifications are necessarily restricted.

LOAD DEFLECTION

The difference in position and orientation of a coupling device between a non-loaded condition and an externally loaded condition. Either or both static and dynamic (inertial) loads may be considered.

MANIPULATOR

Manually controlled motion device used mainly for material handling tasks.

MOBILE ROBOT

A robot mounted on a movable platform.

MOTION SPACE

The volume swept by the motions of the arm components.

PASSIVE COMPLIANCE

Compliant behaviour of the end effector in response to forces exerted on it. No sensors, controls or actuators are involved. In a co-ordinate system this is provided by remote centre compliance acting at the tip of a gripped part.

PICK AND PLACE UNIT

Automatic motion device motions related to sequential action and/or translational or rotational movement are executed according to a fixed programme which cannot be modified without physical intervention. Pick and place units are usually equipped with grippers and are mainly used for material-handling tasks.

POINT-TO-POINT CONTROL

A type of robot control in which the movement is defined by a sequence of points through which the robot must pass (see also annex V).

PRISMATIC JOINT (commonly called "sliding joint")

An assembly between two rigid members enabling one to have a linear motion in contact with the other.

RESOLUTION

The minimum interval between two adjacent discrete details distinguishable one from the other.

REACHABLE SPACE

The volume inside which any point can be reached by the effector coupling device.

RELIABILITY

The probability of an industrial robot performing a specified task without failure in specified operating conditions over a specified time period.

REPEATABILITY

The closeness of agreement between successive results obtained when a specified operation is performed a specified number of times at one set-up. It may be expressed as the error range with a probability of 0.95 for a specified number of measurements.

ROTARY JOINT

An assembly connecting two rigid members which enables one to rotate in relation to the other about a fixed axis.

SETTLING LINE

The time required for a damped oscillatory response to reach the specified limit.

STABILITY

This characteristic concerns the end effector oscillations around a fixed point while the servo-motors are actuated.

STANDARD CYCLE

A sequence of movements made by a robot during a typical task taken as a reference; the cycle description should include the load and the duration of maximum speed and acceleration paths.

TELEOPERATOR

Remote-controlled manipulator

WRIST

An end part of the robot consisting of rotary joints which orient the end effector.

<u>Annex V</u>

INTERNATIONAL ORGANIZATION FOR STANDARDIZATION (ISO)

<u>Classification of industrial robots (draft)</u>

Industrial robots are classified by their:

- Main power source

- Motion control

- Programming method

- Sensory interaction

1. <u>Classification by main (actuating) power source</u>

The main power sources are broken down into the following main types:

1.1 Pneumatic

1.2 Hydraulic

1.3 Electric

2. <u>Classification by motion control system</u>

The motion control systems are broken down into the following main types:

2.1   <u>Point-to-point control (PTP)</u>

2.1.1 Mechanical point-to-point control:

- Each controlled motion operates between mechanically adjustable positions

- The movements in the different axes are not co-ordinated with each other and may be executed simultaneously or consecutively

- Velocities are not specified by the input data

2.1.2 Continuous point-to-point control:

- Each controlled motion operates in accordance with instructions which specify only the next required position

- The movements in the different axes are not co-ordinated with each other and may be executed simultaneously or consecutively

- Velocities are not specified by the input data

2.1.3 Continuous point-to-point with velocity control:

- Each controlled motion operates in accordance with instructions which specify both the next required position and the required velocity to that position

- The controlled motions in the different axes are not co-ordinated with each other

- The controlled motions take place parallel to the machine axes

2.2   Continuous path control (CP)

- Two or more controlled motions operate in accordance with instructions that specify the next required position and the velocity required to reach the positions

- The velocities are varied in relation to each other so that a desired contour is generated

3.   Classification by programming methods

The programming methods are broken down into the following main types:

3.1   Variable sequence programming and mechanical adjustment of the axes positions

3.2   Teach-in programming

3.2.1   Teach-in by manually leading the robot arm

3.2.2   Teach-in by leading a simulating device

3.2.3   Teach-in by using a manually controlled programming unit

3.3   Off-line programming

3.3.1   Explicit or analytic programming

3.3.2   Goal-directed programming

4.   Classification by type of sensory interaction

The sensory interactions are broken down into the following main types:

4.1   No sensing

The robot's interaction with the external environment is limited to basic programme start/stop and emergency stop logical signals.

4.2   On-off Sensing

The robot can sense on-off conditions in the external environment such as from switch contact closures or light-beam interruption and use the information for synchronization with an external process as in a HOLD function, for branching to different stored programme sequences or to terminate a programmed motion as in a search function.

4.3 Continuous adaptive sensing

Robot axis motion is referenced to an external co-ordinate system
as measured by a single dimension sensor for example in a line-
tracking function.

4.4 Multidata adaptive sensing

More flexible and versatile robot operation is attainable through
increased levels of sensory interaction. This includes, for example,
force, touch, proximity or vision. The system is characterized by
several dimension inputs, a high degree of data handling and
interaction with the control system.

Annex VI

PRODUCTION AND USE OF INDUSTRIAL ROBOTS IN SWEDEN IN 1982[1]

Introduction

Industrial robots (IRB) are part of a technology which has rapidly become an important tool for increasing productivity in industry. Together with that of digital computers and computer-aided manufacturing equipment (CNC-machines, computer-controlled material handling systems, CAD/CAM, etc.) their use has transformed manufacturing technology. The most important shift to date in the history of manufacturing occurred when steam engines and electric motors replaced human force. The main characteristic of the current and perhaps equally important shift is the increasing automation of the information flow, as a result of which the manufacturing process will tend to be organized more and more according to continuous-flow principles similar to the closed continuous systems used in such processing industries as the iron and steel, and pulp and paper industries.

Although the industrial-robot industry dates back to the end of the 1960s it is still in its infancy. As may be seen below, the industrial-robot industry, when compared, for instance, with the machine-tool industry, is a marginal industry. However, it is important - one might even say strategic - because

- It is an industry with a very high growth potential; and

- It has become a symbol for factory automation in general and thus has an indirect influence on the diffusion of computer-aided design/computer-aided manufacturing (CAD/CAM)

Definition, classification and statistical data

The present Note takes strictly into account the definition of industrial robots proposed by ECE and ISO 2/ as well as the prerequisite of reprogrammability by means of software. This means that the following equipment is not included in the present study: loading/unloading equipment attached to machine-tools (fixed sequence of operations); automatic manipulators reprogrammable only by means of hardware; and special computer-controlled machinery such as computer-controlled assembly automates.

Industrial robots can be broken down into subgroups according to characteristics such as the drive system, degrees of freedom, load capacity,

---

1/ Submitted by the Government of Sweden. The part of this contribution containing two examples of investment calculations for industrial-robot installations may be found in chapter VIII of the present study.

2/ See Chapter III of the present study. This definition is very close to that proposed by the Robot Institute of America.

sensory perception, price, motion control, operational modes and so on. For the purpose of the present study, industrial robots have been broken down into two main groups: 3/

- Programmable robots; and

- Robots with sensory perception

However, taking into consideration available empirical data, it may be more correct to speak of "advanced robots" (with respect to control system, motion control, programming methods etc.) and "less advanced robots". The robots in the first group generally cost from $US 10,000 to 50,000 and those in the second in the order of $US 50,000.

Each of the two subgroups - programmable robots and robots with sensory perception - have been further subdivided with respect to:

(a) <u>Operational modes</u>

- Handling robots (workpiece gripped by robot)

- Process robots (tool handled by robot)

- Welding, glueing and other joining applications
- Spray-painting
- Deburring, polishing, cutting and other metalworking
- Other process applications (mainly inspection)

- Assembly robots.

(b) <u>Application fields</u> (sectors, branches)

- Engineering industries

- Motor vehicles (including subcontractors)
- Electrical and electronic engineering
- Other engineering

- Other industrial sectors, excluding engineering

- Non-industrial sectors (mainly R and D and education).

In the present Note statistics on production, export, import and foreign trade are given in terms both of number of units and of value in millions of Swedish kronor. The data expressed in value terms include the values of peripheral units (fixtures, welding guns, camera systems etc.) where these have been supplied by the robot vendors.

3/ The adoption of this breakdown was proposed by the Working Party on Engineering Industries and Automation.

It should be noted that the value of imports has been overestimated because it includes not only the value of the imported robot units (which is the "true" import) but also, to a large extent, peripheral units that are domestically produced.

Production, foreign trade and domestic supply

During 1982 the production of industrial robots in Sweden amounted to 912 units, or, in value, to SKr 376 million (tables VI.1 and VI.2).

Table VI.1. Production, export, import and domestic supply
of industrial robots in Sweden (1982)

(Units)

Type of robot	Production	Export (export ratio)	Import	Domestic supply a/
Programmable robots	160 (18 %)	101 (63 %)	31	90 (42 %)
Robots with sensory perception	752 (82 %)	704 (94 %)	77	125 (58 %)
Total	912 (100 %)	805 (88 %)	108	215 (100 %)

Source: Computers and Electronics Commission (DEK)

a/ Domestic supply = production - export + import.

The production of robots with sensory perception accounts for 82 per cent in terms of units and 95 per cent in terms of value of total IRB production, pointing to the fact that the Swedish robot industry is strongly committed to the production of intelligent and advanced robots.

The Swedish robot industry is directed mainly towards international markets. As shown in table VI.1 above, the export ratio in 1982 amounted to 88 per cent in terms of units and 92 per cent in terms of value. The export ratio for robots with sensory perception is considerably higher than that for programmable robots. Of the total number of units exported, 97 per cent are produced in Sweden, and 3 per cent have been imported and then re-exported, mainly to Nordic countries.

In 1982, Sweden imported 108 robots, of which 24 were subsequently re-exported. Thus, of the 215 robots that were supplied to the Swedish market, 84 were imported (39 per cent) and 131 (61 per cent of the total domestic supply) were of domestic production.

Table VI.2. **Production, export, import and domestic supply
of industrial robots in Sweden (1982)**

(Value in millions of Swedish kronor) a/

Type of robot	Production	Export (export ratio)	Import	Domestic supply b/
Programmable robots	20 (5 %)	14 (79 %)	13	19 (29 %)
Robots with sensory perception	356 (95 %)	333 (94 %)	23	46 (71 %)
Total	376 (100 %)	347 (92 %)	36	65 (100 %)

Source: Computers and Electronics Commission (DEK)

a/ $US 1 = SKr 7.50

b/ Domestic supply = production - export + import.

With respect to operational modes, IRB can be broken down into the
following three groups:

- Workpiece gripped by robot, handling robots. Material handling,
  loading/unloading machines (variable sequence of operations) etc.;

- Tool handled by robot, process robots. Welding, glueing, deburring,
  cutting, painting, polishing, inspection etc.; and

- Assembly, assembly robots.

Of the robots produced in Sweden in 1982, handling robots accounted for
32 per cent (14 per cent in terms of value) and process robots for 68 per cent
(86 per cent in terms of value). There was no production of assembly robots
(table VI.3).

With respect to operational modes, there is a significant difference between
programmable robots and robots with sensory perception. Programmable robots are
used mainly as handling robots (88 per cent in terms of units and 85 per cent in
terms of value) (table VI.3). Robots with sensory perception, on the other hand,
are used mainly as process robots (80 per cent in terms of units and 90 per cent
in terms of value).

Table VI.3. Production of industrial robots in Sweden (1982)

(Distribution by operational mode)

Operational mode	Programmable robots		Robots with sensory perception		Total	
	Units	Value[a]	Units	Value[a]	Units	Value[a]
Handling robots	140 (88 %)	17 (85 %)	152 (20 %)	36 (10 %)	292 (32 %)	53 (14 %)
Process robots	20 (12 %)	3 (15 %)	600 (80 %)	320 (90 %)	620 (68 %)	323 (86 %)
Total	160 (100 %)	20 (100 %)	752 (100 %)	365 (100 %)	912 (100 %)	376 (100 %)

Source: Computers and Electronics Commission (DEK).

a/ In millions of Swedish kronor; $US 1 = SKr 7.50.

Process robots are used mostly for joining operations (mainly glueing and welding),
53 per cent in terms of units of total production and 41 per cent in terms of
value being employed for this purpose (see table VI.8). In that table, the robots
grouped under "other" (i.e. inspection) account for only 6 per cent of the total
number of robots produced but for 39 per cent in terms of value. This large
difference between unit share and value share stems from the fact that the robots
in this group are equipped with very expensive peripherals, i.e. cameras and
picture processing units.

Domestic supply during 1982 shows by and large the same picture as production,
that is, the domination of process robots (table VI.4). However, handling robots
have a significantly larger share in domestic supply than in production,
accounting for 43 per cent (in terms of units) of domestic supply. The equivalent
share for production is 32 per cent.

Table VI.4. Domestic supply of industrial robots in Sweden (1982)

(Distribution by operational mode)

Operational mode	Programmable robots		Robots with sensory perception		Total	
	Units	Value[a]	Units	Value[a]	Units	Value[a]
Handling robots	62 (69 %)	11 (58 %)	30 (24 %)	13 (28 %)	92 (43 %)	23 (35 %)
Process robots	28 (31 %)	8 (42 %)	90 (72 %)	31 (67 %)	118 (55 %)	40 (62 %)
Assembly robots	0	0	5 (4 %)	2 (4 %)	5 (2 %)	2 (3 %)
Total	90 (100 %)	19 (100 %)	125 (100 %)	46 (100 %)	215 (100 %)	65 (100 %)

Source: Computers and Electronics Commission (DEK).

a/ In millions of Swedish kronor; $US 1 = SKr 7.50.

It may be seen in table VI.4 that assembly robots account for only a marginal
share of total domestic supply: 2 per cent in units and 3 per cent in value.
All assembly robots put on the Swedish market during 1982 were imported.

Distribution of robots by application field (sector)

     The engineering industries are the main users of IRB. Their share of both
production and domestic supply in 1982 was 86 per cent in terms of units
(84 per cent and 88 per cent, respectively, in terms of value) (tables VI.9-11).

Other industrial sectors accounted for 10 per cent of production in 1982
(9 per cent in value terms). Domestic supply showed roughly the same shares.
In these sectors, programmable robots had a larger share than robots with sensory
perception, 25 per cent of the production (in units) of the former going to other
industrial sectors and only 7 per cent of the latter. Non industrial sectors,
mainly R and D and education, accounted for 4 per cent of production in terms of
units (7 per cent in value terms). All robots supplied to this sector in 1982
were robots with sensory perception.

     Among subsectors in the engineering industries the predominant user is the
motor-vehicle industry (including subcontractors). Of total robot production,
this subsector accounted for 72 per cent in terms of units and 77 per cent in
terms of value (table VI.11). This table also shows the significant difference
between production and domestic supply, the corresponding figures for domestic
supply being considerably lower - 44 per cent in terms of units and 45 per cent
in terms of value. The major reason for this difference is that the Swedish
automotive industry started to invest in robots earlier than automotive
industries in other countries. The heavy investments in spot-welding lines, for
example, have already to a large extent been completed in Sweden.

Total robot population at the end of 1982

     The total IRB population in Sweden at the end of 1982 can be estimated at
between 1,600 and 1,700 units. Table VI.5 shows the diffusion of IRB since 1970.

Table VI.5. The diffusion of industrial robots in Sweden (1970-1984)

	1970	1973	1977	1979	1981 (September)	1982	1984 (forecast)
Total robot population	55	135	490	940	1 250	1 600– 1 700	2 300

     Source: Computers and Electronics Commission (DEK).

With 1,600 to 1,700 robots, Sweden has the highest robot density in the world.
According to the OECD study, "The Impact of Industrial Robots on the Manufacturing
Industries of Member Countries", the number of robots per 10,000 persons employed
in manufacturing in 1981 was 30 in Sweden, 13 in Japan, 5 in the Federal Republic
of Germany and 4 in the United States.

<u>The Swedish robot industry (size and structure)</u>

Taking a broad definition of industrial robots, manufacturers can be broken down into the following three categories:

(a) <u>Manufacturers of general purpose IRB</u>.

(b) <u>Manufacturers of special purpose IRB</u>, i.e. automatic loading/unloading equipment attached to specific machines. Volvo's "Doppin" for loading/unloading press machines is an example of special purpose IRB. Several hundred units of this robot have been sold to automotive manufacturers all over the world.

(c) <u>Manufacturers of programmable material-handling equipment</u>, i.e. auto-carrier systems, computer-controlled crane and warehousing systems. The leading companies in this group are Volvo ACS, BT Lifters, Tellus, Digitron, ASEA and Moving. The robots produced by these manufacturers are essential components of flexible manufacturing systems.

In the group of manufacturers of general purpose IRB - the subject of the present study - the main manufacturers in Sweden are ASEA, Atlas-Copco Tools and Satt-Kaufeldt. Besides these companies, there are some 5 to 10 other smaller manufacturers producing robots either for internal use or for external sale. ASEA, which is one of the world's leading robot manufacturers, accounts for more than 80 per cent of Swedish IRB production. In the class of advanced robots, the market share for ASEA in Europe and the United States was 36 per cent and 7 per cent, respectively, in 1982. The strong penetration of ASEA in international markets has motivated the company to set up production facilities in the United States, Spain, France and Japan. ASEA Robotics now employs about 600 persons, of whom 350 in Sweden.

The major foreign suppliers on the Swedish market are Unimation, Cincinnati Milacron, Trallfa and Yaskawa.

<u>The importance of the robot industry versus the machine-tool industry</u>

While world machine-tool production in 1982 has been estimated at $US 22.7 billion, <u>4/</u> the output of the world industrial robot industry can be estimated in the order of $US 600 million (the sum of the output of the major robot producing countries). The size of the IRB industry is therefore less than 3 per cent of that of the machine-tool industry. Thus, as an industry it still has only marginal importance, but it should be kept in mind that, as mentioned above, the industry is only in its infancy.

However, when comparing the Swedish IRB industry with the country's machine-tool industry, the picture is somewhat different. In 1982 the output in the robot industry amounted to 27 per cent of the output in the machine-tool industry. As the robot industry has a much faster growth rate than the machine-tool industry, this share may increase to 35-40 per cent in 1983. The Swedish IRB industry accounts for about 9 per cent of total world IRB output. This can be compared to the Swedish machine-tool industry which accounts for 1 per cent of world output (roughly the same as the share of the total Swedish economy in the world economy).

<u>4/</u> <u>Production and use of industrial robots in Sweden in 1982</u> Computers and Electronics Commission, Ministry of Industry, DsI 1983:1, Stockholm 1983.

It may thus be concluded that, while the world IRB industry is still marginal in comparison with the machine-tool industry, in Sweden, it is of considerably larger importance. In relation to the domestic machine-tool industry, the IRB industry of Sweden is significantly larger than those of Japan, the United States, the Federal Republic of Germany and many other countries, as illustrated by the figures below:

World IRB output/world machine-tool output	3 %
Swedish IRB output/Swedish machine-tool output (1982)	27 %
(1983)	35-40 %
Swedish IRB output/world IRB output	9 %
Swedish machine-tool output/world machine-tool output	1 %

The following tables were prepared in answer to two questionnaires for the study circulated by the ECE secretariat to member countries. 5/

Table VI.6.	Distribution of robots by operational mode - programmable robots
Table VI.7.	Distribution of robots by operational mode - robots with sensory perception
Table VI.8.	Distribution of robots by operational mode - programmable robots and robots with sensory perception
Table VI.9.	Distribution of robots by application field - programmable robots
Table VI.10.	Distribution of robots by application field - robots with sensory perception
Table VI.11.	Distribution of robots by application field - programmable robots and robots with sensory perception

---

5/ See annexes I and II.

Table VI.6.   Distribution of robots by operational mode – Programmable robots (1982)

Operational mode	Number of units, 1982				Value (millions of Swedish kronor)[a]			
	Production	Export	Import	Domestic supply b/	Production	Export	Import	Domestic supply b/
Workpiece gripped by robot, subtotal (handling-robots)	140 (88%)	88	10	62 (69%)	17 (85%)	9	3	11 (58%)
Tool handled by robot, subtotal (process-robots)	20 (12%)	13	21	28 (31%)	3 (15%)	5	10	8 (42%)
Assembly operations	0	0	0	0	0	0	0	0
Total	160 (100%)	101	31	90 (100%)	20 (100%)	14	13	19 (100%)
Export ratio c/	63%				70%			

Source: Computers and Electronics Commission (DEK).

a/ $US 1 = SKr 7.60.

b/   Domestic supply = Production - export + import.

c/   Export ratio = $\dfrac{\text{export}}{\text{production}}$. 100.

Table VI.7. Distribution of robots by operational mode – Robots with sensory perception (1982)

Operational mode	Number of units, 1982				Value (millions of Swedish kronor)			
	Production	Export	Import	Domestic supply a/	Production	Export	Import	Domestic supply a/
Workpiece gripped by robot, subtotal (handling-robots)	152 (20%)	143	21	30 (24%)	36 (10%)	31	8	13 (28%)
Tool handled by robot, subtotal (process-robots)	600 (80%)	561	51	90 (72%)	320 (90%)	302	13	31 (67%)
Assembly operations	0	0	5	5 (4%)	0	0	2	2 (4%)
Total	752 (100%)	704	77	125 (100%)	356 (100%)	333	23	46 (100%)
Export ratio b/	94%				94%			

Source: Computers and Electronics Commission (DEK).

a/ Domestic supply = Production – export + import.

b/ Export ratio = export / production . 100.

Table VI.8. Distribution of robots by operational mode - Programmable
robots and robots with sensory perception (1982)

Operational mode	Number of units, 1982				Value (millions of Swedish kronor)			
	Production	Export	Import	Domestic supply a/	Production	Export	Import	Domestic supply a/
Workpiece gripped by robot, subtotal	292 (32%)	231	31	92 (43%)	53 (14%)	40	10	23 (35%)
- Joining including welding and glueing	485 (53%)	456	55	84 (39%)	155 (41%)	147	15	23 (35%)
- Surface treatment, mainly spray-painting	20 (2%)	13	17	24 (11%)	3 (1%)	5	9	7 (11%)
- Other metal-working (deburring, polishing, etc.)	60 (7%)	55	0	5 (2%)	20 (5%)	15	0	5 (8%)
- Other (i.e. inspection)	55 (6%)	50	0	5 (2%)	145 (39%)	140	0	5 (8%)
Tool handled by robot, subtotal	620 (68%)	574	72	118 (55%)	323 (86%)	307	24	40 (62%)
Assembly	0	0	5	5 (2%)	0	0	2	2 (3%)
Total	912 (100%)	805	108	215 (100%)	376 (100%)	347	36	65 (100%)
Export ratio b/	88%				93%			

Source: Computers and Electronics Commission (DEK).

a/ Domestic supply = Production - export + import.

b/ Export ratio = $\dfrac{\text{export}}{\text{production}}$ - 100.

Table VI.9. Distribution of robots by application field – Programmable robots (1982)

Sector	Number of units				Value (millions of Swedish kronor)			
	Production	Export	Import	Domestic supply a/	Production	Export	Import	Domestic supply a/
Engineering (ISIC 38)	120 (75%)	75	31	76 (84%)	15 (75%)	11	13	17 (89%)
Other industrial excluding engineering (ISIC 2+3+4-38)	40 (25%)	26	0	14 (16%)	5 (25%)	3	0	2 (11%)
Non-industrial (Mainly R and D and education	0	0	0	0	0	0	0	0
Total	160 (100%)	101	31	90 (100%)	20 (100%)	14	13	19 (100%)

Source: Computers and Electronics Commission (DEK).

a/ Domestic supply = Production – export + import.

Table VI.10.   Distribution of robots by application fields – Robots with sensory perception (1982)

Sector	Number of units, 1982				Value (millions of Swedish kronor)			
	Production	Export	Import	Domestic supply a/	Production	Export	Import	Domestic supply a/
Engineering (ISIC 38)	667 (89%)	624	65	108 (86%)	301 (85%)	281	20	40 (87%)
Other industrial excluding engineering (ISIC 2+3+4-38)	50 (7%)	50	9	9 (7%)	30 (8%)	29	2	3 (7%)
Non-industrial (Mainly R and D and education)	35 (4%)	30	3	8 (7%)	25 (7%)	23	1	3 (6%)
Total	752 (100%)	704	77	125 (100%)	356 (100%)	333	23	46 (100%)

Source: Computers and Electronics Commission (DEK).

a/ Domestic supply = Production – export + import.

Table VI.11. Distribution of robots by application field – Programmable robots and robots with sensory perception (1982)

Sector	Number of units, 1982				Value (millions of Swedish kronor)			
	Production	Export	Import	Domestic supply a/	Production	Export	Import	Domestic supply a/
– Motor vehicles (including subcontractors)	658 (72%)	600	36	94 (44%)	288 (77%)	271	12	29 (45%)
– Electrical and electronic engineering	15 (2%)	13	7	9 (4%)	5 (1%)	4	3	4 (6%)
– Other engineering	114 (12%)	86	53	81 (38%)	23 (6%)	17	18	24 (37%)
Engineering, subtotal	787 (86%)	699	96	184 (86%)	316 (84%)	292	33	57 (88%)
Other industrial excluding engineering b/	90 (10%)	76	9	23 (11%)	34 (9%)	31	2	5 (8%)
Non-industrial (Mainly R and D and education)	35 (4%)	30	3	8 (3%)	25 (7%)	23	1	3 (4%)
Total	912 (100%)	805	108	215 (100%)	375 (100%)	346	36	65 (100%)

Source: Computers and Electronics Commission (DEK).

a/ Domestic supply = Production – export + import.

b/ Chemical and plastic industry: production 20 units, export 9 units, import 3 units, domestic supply 14 units.

Annex VII

MULTILATERAL CO-OPERATION AMONG CMEA MEMBER COUNTRIES IN THE FIELD OF ROBOT TECHNOLOGY 1/

In recent years, CMEA member countries have been devoting considerable attention to the development of industrial robots and their use in production, robot manufacture being one of the most advanced fields of engineering. Given the shortage of labour in the European CMEA member countries, the broad utilization of industrial robots makes it possible to further increase output by automating technological and production processes, particularly auxiliary operations. The use of industrial robots at enterprises engaged in large-scale production is of particular importance, since it releases men from low-skilled, physically hard, monotonous and exhausting work and from the need to work in a harmful environment; labour resources can then be transferred to other sectors of production in a planned fashion.

At present, CMEA member countries produce more than 150 different types of industrial robots. These robots are used most widely in servicing metal-cutting tools, press-forging plants and casting equipment and in welding, painting, assembly, electroplating, transport and storage operations, where they can carry out operations beyond the physical abilities of man. Later on, as robot construction develops in CMEA member countries, industrial robots will be used in other branches of the national economy, for example in the coal industry, metallurgy, the radio engineering and electronics industries, light industry and the food industry.

The economic impact of industrial robots is significant. Experience shows that the problems associated with the use of industrial robots must be resolved in an integrated manner and that the effective use of industrial manipulator robots in individual production processes requires a restructuring of the technology used in these processes, i.e. the development of flexible and easily reset systems and the utilization of modern methods of labour organization and new production management methods.

All these considerations are taken into account when activities are being co-ordinated within the framework of co-operation among CMEA countries in the field of industrial robot technology. In the last three years, measures have been taken within the CMEA framework to pool efforts and co-ordinate activities in the field of robot technology. In 1979, CMEA member countries, operating within the framework of the CMEA Standing Commission on Co-operation in the Field of Engineering, carried out a technical and economic analysis and prepared a long-term forecast of the development of industrial robots and manipulators. Their aim was to select areas of scientific and technological co-operation among CMEA member countries and to promote specialization and collaboration in the production of promising types of industrial robots and manipulators with a view to meeting the related requirements of CMEA member countries.

1/ Transmitted by the Secretariat of the Council for Mutual Economic Assistance (CMEA), Moscow. The active co-operation between ECE and the CMEA Secretariat in the field of robotics has also been described in: E. Zador, I. Botchkarev, Z. Vacek, "Robots are coming". Press Bulletin, No. 108, CMEA Moscow 1984.

In 1980, upon the recommendation of CMEA, the relevant organizations of CMEA member countries signed the Agreement on Scientific and Technological Co-operation for the Development of High-Potential Designs for Multipurpose Industrial Manipulator Robots (primarily the development of new robots for metal-working and casting equipment, technological operations, assembly, welding and painting). The programme of co-operation within the framework of the aforementioned Agreement involves 27 subject areas, one-third of which have already been the subject of research aimed at improving the design of industrial robots developed earlier in CMEA member countries and developing new models. The organizations participating in this co-operation have jointly developed standard multipurpose grippers and technical materials selected in the light of the range of available industrial robots and their applications. They have also standardized components and assemblies and established technical requirements for these components and assemblies, in particular for electric drives, hydro and pneumatic equipment for the drives, programme control devices, sensors and systems for the adaptation of industrial robots. Methodological manuals have been compiled for designers and industrial engineers engaged in the automation of production and in the introduction of industrial manipulator robots in production processes. These manuals concern, in particular:

> Grippers for industrial robots, their classification, degree of automation, principle of operation, gripping force, number of positions, replacement methods, lifting-capacity standards and maximum dimensions of handled parts;

> Assembly tools for industrial robots, with a priority list of assembly tools for standard assembly operations, and a description of the design and operating principles of various types of assembly tools;

> Improvement of industrial robot designs and their use as elements in technological modules, due account being taken of laboratory and industrial tests;

> Selection of range of basic technological equipment for "equipment/robot" technological modules and other materials.

A catalogue entitled "Industrial Robots" has been compiled and published. It includes 129 industrial robot modules, describing their technical specifications and making recommendations with regard to robot-serviced equipment.

On the basis of the aforementioned scientific and technological studies in the field of robot technology conducted in CMEA member countries in the period prior to 1981, and in the light of the importance of the use of industrial manipulator robots for the national economy, the heads of Government of the CMEA member countries, acting in accordance with a decision taken by the Governments of CMEA member countries at the thirty-sixth meeting of the Session of the Council for Mutual Economic Assistance, in 1982 signed the General Agreement on Multilateral Co-operation for the Development and Organization of Specialized Co-production of Industrial Robots. This General Agreement served as the basis for the establishment of the Board of Chief Designers in Industrial Robot Technology within the framework of CMEA. The Board's primary task is to formulate concepts for the technical development of robot technology within the framework of long-term co-operation, prepare proposals for the realization of these concepts and supervise co-operation among CMEA member countries in developing industrial robots on the basis of standardized modules, assemblies and components. The Board has now drawn up and approved a work entitled "Concept of the technical development of robot technology in the context of the organization of co-operation among CMEA member countries", which sets out the main objectives of the technical development of industrial robot technology in order to secure reciprocal deliveries of standardized assemblies and components, auxiliary devices and tools for manufacturing robots for the various branches of the national economy.

The use of the modular principle for the manufacture of industrial robots from standardized modules, assemblies and parts will make it possible to:

Reduce the time required for robot design, manufacture and entry into use;

Cut down the production cost of industrial robots;

Increase their operational reliability through the use of advanced designs;

Facilitate operation and repair as a result of the reduction in the number of elements used.

One of the main areas for the development and introduction of industrial robots is their use for the automation of industrial processes. In this connection, in the CMEA member countries it is planned to use industrial robot technology to develop flexible automated production systems requiring minimum human inputs. In line with the General Agreement, agreement has been reached on a Programme for Co-operation among CMEA Member Countries on the development of Robot Systems. This Programme comprises the following main elements:

Robot systems for processing components such as shafts and flanges;

Robot systems for hot and cold forging and casting, assembly, control and measuring operations.

At the beginning of 1983, organizations of CMEA member countries concluded an Agreement on Multilateral International Specialization and Collaboration in the Production of Industrial Robots. The Agreement envisages further work in CMEA member countries on the production of industrial robots, including the manufacture of 57 types of standard industrial robtos designed for various purposes. This represents the first stage in the multilateral co-operation of CMEA member countries in the field of the specialized production of industrial manipulator robots. The results of co-operation in this field will serve as the basis for the further division of labour in robotics and also for the broader application for this type of modern equipment in the various branches of the national economies of CMEA member countries.

<u>Annex VIII</u>

PROSPECTS FOR THE DEVELOPMENT AND USE OF
INDUSTRIAL ROBOTS IN THE USSR 1/

The need to increase labour productivity and save labour resources means that industry must develop fundamentally new mechanisms for automating working processes, mechanisms suited to conditions involving frequent changes in the product being manufactured. These mechanisms are referred to as automatic programmable-control manipulators (industrial robots). The main features of the use of industrial robots are as follows:

Increased product quality and output using the same number of workers (or fewer workers); this is possible thanks to the shorter time required for each operation, the maintenance of a constant "fatigue-free" work rate, the greater flexibility of equipment, the improvement of existing high-speed processes and equipment and the creation of incentives for the development of new such processes and equipment;

New working conditions, with the release of workers from non-skilled, monotonous, heavy and dangerous work, the improvement of safety and the reduction of the amount of time lost as a result of industrial accidents and work-related illnesses;

Economy of labour, release of workers for other national economic tasks and the fulfilment of the social commitment to release the great majority of workers for creative and skilled work.

All industrial robots can be divided into three groups on the basis of the operations they perform:

Hoisting-and-carrying industrial robots able to "lift, carry and place". These robots are used to service main production units through auxiliary operations involving positioning and removing semis, parts, tools and attachments, cleaning bases, components and equipment, feeding conveyors and carrying out handling, storage and other similar operations;

Production robots which perform the main operations in a technological process, for example assembling, welding, painting, bending and cutting;

Universal industrial robots intended for a broad range of uses and capable of performing varied technological operations, both main operations and auxiliary operations.

1/   Transmitted by the Government of the Union of Soviet Socialist Republics.

The tables included in this contribution have been incorporated into the text of the study (chapters III, IV, V), as have most of the figures, which were redrawn by the United Nations services (see figs. 4-9).

More than 250 firms and organizations are currently involved in developing
and manufacturing industrial robots. The number of known models of industrial
robots manufactured throughout the world exceeds 500. It can be said that the
world's industrialized countries are now witnessing the rapid growth of a new
industrial sector, namely robot manufacture. The main application for industrial
robots is to service press-forging equipment, casting machinery and metal-cutting
tools. However, the use of industrial robots for welding, painting and the
application of protective coatings continues to expand. In addition, a number
of assembly operations have recently been automated with the aid of industrial
robots.

Industrial robots are used in different types of production:

In mass production and large-scale production, they are used as fixed and
flexible links for automated lines. The use of industrial robots precludes the
need to develop special means of handling and transport in each specific case
and makes it possible to reset automated lines, set them up and bring them into
service rapidly;

In series production involving a range of products, industrial robots are
used with resettable automated lines consisting of standard-unit machine tools
and adjustable lines designed for batch component processing, including automated
lines with programmable control;

In small-scale production, industrial robots are used for the automated
loading of equipment forming automated units. The use of industrial robots
facilitates automation regardless of whether the unit is organized along
functional lines (single type of equipment) or production-process lines
(ensuring required processing route).

An analysis of 500 models makes it possible to identify a number of features
characteristic of modern industrial robot designs. Most industrial robots use a
cylindrical system of co-ordinates, although the number of designs based on
angular co-ordinates has recently increased. The most common industrial robots
have four or five degrees of freedom. The main types of drives used are
hydraulic and pneumatic, although a trend has recently developed towards the
wider use of electric drives: in 1975, of the total number of 300 models,
4.5 per cent were driven electrically, whereas the figure is now 13.5 per cent.
The load carrying capacity of industrial robots varies widely, but most models
(65 per cent of total) are designed to carry between 10 and 40 kg. As many
as 60 per cent of industrial robots are now equipped with programmable
positioning systems with a memory able to handle from 10 to 50 commands; they
have up to 10 channels for external linkages.

The positioning error of industrial robots of known designs lies between
±0.02 and ±5.0 mm. In 70 per cent of cases, it is less than ±1 mm. The
dimensions of industrial robots, which determine the amount of space they occupy
(a complex indicator which also takes account of working area) also varies
considerably. At present, most industrial robots are simple in design, having a
positioning system with programmable control and a single arm with up to
four degrees of freedom.

With regard to the development of industrial robot designs, considerable
efforts are being made to develop economical systems involving the construction
of specialized and universal models from standard modules. The work being carried
out in this connection in Bulgaria, Czechoslovakia, the Federal Republic of
Germany, Japan, the USSR, the United States of America and a number of other

countries shows that modular construction makes it possible to develop specialized machinery that more fully satisfies the needs of the consumer, makes the design, manufacture and introduction of industrial robots cheaper and quicker and makes the robots easier to repair and set up. Work going on in CMEA member countries to develop national systems for the modular construction of industrial robots will probably be completed by the end of 1985. The success of such work will depend on the creation of a unified system of standardized dimensions for industrial robots and of CMEA standards governing robot terminology and definitions, establishing their basic technical specifications, dealing with questions of structure and classification and laying down rules in respect of safety techniques, testing methods and approval procedures. The completion of the research that has been agreed upon, the development of industrial robot designs and the issuance of basic CMEA standards will help raise the technical standard of robots, reduce their cost, permit assembly on the basis of jointly produced components and reduce the amount of labour required to couple robots to main technological units as a result of the development of uniform technical specifications based on common technical requirements. The use of modular construction methods for industrial robots will also facilitate the development of robots with a large kinematic potential, developed sensory apparatus and appropriate algorithmic programmes. Industrial robots of this type will be used extensively for the selection and recognition of components for the purposes of assembly, packing, painting, welding, etc.

An analysis of work on the development of modular industrial robots and the technical parameters of contemporary models allows several preliminary recommendations to be made in connection with the development of similar robots. The main goal must be to develop complete units (modules) designed to carry out specific functions, for example a set of control system modules (control station, electrical station, set of sensors for information subsystem), a set of kinematic modules for robot mechanical systems (units for advancing and retreating, turning, lifting and turning, etc.), a set of grippers and a power plant. Every effort should be made to ensure that the coupling elements (outlets) of modules designed for a given function are standardized so that, for example, modifications can be made to the mechanical systems of industrial robots using different programme control systems.

Work is continuing with a view to improving industrial robot design (for example by extending the range of models equipped with electromechanical and pneumatic drives), improving control and automation systems, increasing the use of sensors, etc. Increasing the functional capacity of industrial robots in order to expand their range of applications has become necessary primarily as a result of the need to develop highly automated production systems capable of operating in an all-machine environment and relying less and less on man to operate them and control them.

With regard to casting, work should be completed shortly on improving the design of industrial robots intended to service die-casting equipment and thermoplastic units; this work involves developing reliable designs for pneumatic drives and hydraulic drives operating on non-combustible fluids, increasing operational rapidity through the extensive use of pneumatic drives and multi-arm systems, using industrial robots to carry out auxiliary operations connected with removing gates, checking the presence and condition of semis, etc. Industry needs new types of industrial robots for the automation of sand casting processes, to service moulding machinery (installation and removal of moulding boxes, assembly of moulds), for casting, including conveyor operations, for fetting and checking castings and for packing bulk-stored castings in containers.

With regard to automating the auxiliary operations involved in servicing press-forging equipment, efforts must be pursued to improve industrial robot design with a view, among other things, to making robots faster and more reliable and providing them with adaptation systems (sensors monitoring the presence of semis, their weight and temperature, position at loading point, etc.).

Industry now needs robots which can search for, recognize and correctly set badly positioned components. Such mechanisms should be extremely useful in the automation of assembly operations and at initial work stations in automatic lines and automated factory units. The automation of assembly operations using industrial robots should proceed rapidly in the next few years.

As a result of the trends that have developed, efforts in the USSR and in most other industrially developed countries to increase the available number of industrial robots involve the manufacture of simple robots (intended for simple operations, with up to four degrees of freedom and a memory with a capacity of not more than 100 points), multi-function programmable robots (with four or more degrees of freedom and a developed numerical control system, including systems based on minicomputers and microcomputers) and adaptive robots fitted with developed sensory equipment (see table 42 in chapter V).

Given the continuing efforts being made to improve the design of industrial robots (for example by expanding the range of models fitted with electromechanical drives), to improve control and automation systems and to expand the use of sensors, it should be possible, by 1985-1990, to attain the performances indicated in table 5 (chapter III).

Changes in the distribution of industrial robots by field of application are shown in table 12 (chapter IV). The functions of robots will also change, thanks to the inclusion in robot work programmes of certain operations currently performed by workers.

With regard to expanding the field of application of industrial robots, efforts are being concentrated on the development of "equipment - attachments - robots" sets. Universal robots are already being proposed to the consumer as part of specific technological set-ups for welding lines, hot-stamping installations, die casting units, etc. Increasing attention is being paid to the idea of developing specialized industrial robots able to serve a wide range of equipment models designed for a single technological purpose.

As far as the development of flexible production systems is concerned, development efforts are being concentrated more and more on the integrated development of automated production units, lines and factory shops. In this connection, there is a recognized need to work simultaneously and in a co-ordinated fashion on the development (and use) of technological equipment, the automation of production processes, information systems and control devices, the formulation of appropriate organizational and technical measures and the resolution of socio-economic problems.

Annex IX

ECE publication on engineering industries and automation (as from 1981)

Annual Reviews of Engineering Industries and Automation:

(1) **1979** (ECE/ENG.AUT/3) UN Sales No.: E.81.II.E.16, New York, 1981

(2) **1980** (ECE/ENG.AUT/7) UN Sales No.: E.82.II.E.18, New York, 1982

(3) **1981** (ECE/ENG.AUT/10) UN Sales No.: E.83.II.E.20, New York, 1983

(4) **1982** (ECE/ENG.AUT/17) UN Sales No.: E.84.II.E.12, New York, 1984

Bulletins of Statistics on World Trade in Engineering Products:

(5) **1979**, UN Sales No.: E/F/R.81.II.E.13, New York, 1981

(6) **1980**, UN Sales No.: E/F/R.82.II.E.5, New York, 1982

(7) **1981**, UN Sales No.: E/F/R.83.II.E.8, New York, 1983

(8) **1982**, UN Sales No.: E/F/R.84.II.E.5, New York, 1984

Other techno-economic studies:

(9) **Development of Airborne Equipment to Intensify World Food Production** (ECE/ENG.AUT/4), United Nations Publication, Sales No.: E.81.E.24, New York, 1981

(10) **Techno-Economic Aspects of the International Division of Labour in the Automotive Industry** (ECE/ENG.AUT/11), United Nations Publication, Sales No.: E.83.II.E.14, New York, 1983

(11) **Engineering Equipment and Automation Means for Waste-Water Management in ECE Countries**, Volume I and Volume II (ECE/ENG.AUT/18, Vol.I and Vol.II), United Nations Publication, Sales No.: E.84.II.E.13 and E.84.II.E.23, New York, 1984

(12) **Measures for Improving Engineering Equipment with a View to More Effective Energy Use** (ECE/ENG.AUT/16), United Nations Publication, Sales No.: E.84.II.E.25, New York, 1984

Reports of Seminars held under the auspices of the ECE Working Party on Engineering Industries and Automation as from 1981:

(13) **Seminar on Automation of Assembly in Engineering Industries**, Geneva, Switzerland, 22-25 September 1981 (AUTOMAT/SEM.8/3)

(14) **Seminar on Present Use and Prospects for Precision Measuring Instruments in Engineering Industries**, Dresden, German Democratic Republic, 20-24 September 1982 (ENG.AUT/SEM.1/3)

(15) **Seminar on Innovation in Biomedical Equipment**, Budapest, Hungary, 2-6 May 1983 (ENG.AUT/SEM.2/3)

(16) **Seminar on Flexible Manufacturing Systems: Design and Applications**, Sofia, Bulgaria, 24-28 September 1984 (ENG.AUT/SEM.3/4)